IRELAND, THE GREAT WAR AND THE GEOGRAPHY OF REMEMBRANCE

Nuala C. Johnson explores the complex relationship between social memory and space in the representation of war in Ireland. The Irish experience of the Great War, and its commemoration, is the location of Dr Johnson's sustained and pioneering examination of the development of memorial landscapes, and her study represents a major contribution both to cultural geography and to the historiography of remembrance. Attractively illustrated, this book combines theoretical perspectives with original primary research showing how memory took place in post-1918 Ireland, and the various conflicts and struggles that were both a cause and effect of this process. Of interest to scholars in a number of disciplines, *Ireland, the Great War and the Geography of Remembrance* shows powerfully how Irish efforts to collectively remember the Great War were constantly in dialogue with issues surrounding the national question and how the memorials themselves bore witness to these tensions and ambiguities.

NUALA C. JOHNSON is a Lecturer in Geography at Queen's University, Belfast. She is currently completing a co-edited volume *A Companion to Cultural Geography* (2003).

T0279839

Cambridge Studies in Historical Geography 35

Series editors:
ALAN R. H. BAKER, RICHARD DENNIS, DERYCK HOLDSWORTH

Cambridge Studies in Historical Geography encourages exploration of the philosophies, methodologies and techniques of historical geography and publishes the results of new research within all branches of the subject. It endeavours to secure the marriage of traditional scholarship with innovative approaches to problems and to sources, aiming in this way to provide a focus for the discipline and to contribute towards its development. The series is an international forum for publication in historical geography which also promotes contact with workers in cognate disciplines.

For a full list of titles in the series, please see end of book.

IRELAND, THE GREAT WAR AND THE GEOGRAPHY OF REMEMBRANCE

NUALA C. JOHNSON

The Queen's University of Belfast

CAMBRIDGE
UNIVERSITY PRESS

CAMBRIDGE UNIVERSITY PRESS
Cambridge, New York, Melbourne, Madrid, Cape Town, Singapore, São Paulo

Cambridge University Press
The Edinburgh Building, Cambridge CB2 8RU, UK

Published in the United States of America by Cambridge University Press, New York

www.cambridge.org
Information on this title: www.cambridge.org/9780521826167

First published 2003
This digitally printed version 2007

A catalogue record for this publication is available from the British Library

ISBN 978-0-521-82616-7 hardback
ISBN 978-0-521-03705-1 paperback

Contents

Illustrations

Acknowledgements

At last this book is written. It was a long struggle but made all the more pleasurable by the many colleagues and friends who offered encouragement, enthusiasm, and constructive criticism over the course of this project. There are many more people who contributed to my thinking about memory and the First World War than can be mentioned here. Special thanks are due, however, to the following who willingly engaged in discussions about various aspects of the Great War in general and Irish social memory in particular. I am most especially grateful to Fred Boal, Barry Carroll, Hugh Clout, Julia Cream, Richard Dennis, Jim Duncan, Brian Graham, Mike Heffernan, Mark Hennessy, Peter Jackson, Alun Jones, Gerald Mills, Máirín Nic Eoin and Hugh Prince. My colleagues at Queen's University proved especially helpful in the final stages of completing this work. Thanks to Keith Lilley, Niall Majury, Satish Kumar, Lindsay Proudfoot, James Ryan and Steve Royle for keeping me cheerful and focused on the task at hand.

I owe a debt of gratitude to the librarians and archivists who provided much needed help at the National Library of Ireland; Department of Early Printed Books, Trinity College Dublin; National Archive of Ireland; Imperial War Museum; University College London Library; Special Collections, Queen's University Belfast; National Photographic Archive; and the Public Record Office of Northern Ireland. I also wish to acknowledge the British Academy who provided funding to do part of the research for this book.

I wish to thank the Series Editors, Alan Baker and Deryck Holdsworth, the referees and the staff at Cambridge University Press for offering assistance and encouragement throughout the various stages of the book's production. Special thanks are also due to Maura Pringle and Gill Alexander of Queen's University for producing the maps and illustrations.

The efforts of John Agnew and Charlie Withers in offering invaluable free advice on an earlier draft of this book are most appreciated. The endless insights, thoughtful criticism and constant encouragement provided by my colleague and friend David Livingstone helped to clarify the more murky ideas flowing through my head and the final written text. For all that I am most grateful. Finally, I wish to thank my parents, Thomas and Ann Johnson, for their steady support and interest in this project over the years and it is to them that I dedicate this book.

1

Geography, landscape and memory

On a grey, wet Sunday in October 2001 the bodies of nine men executed and buried in Mountjoy gaol in Dublin were exhumed and reinterred at Glasnevin cemetery.[1] Thousands of people lined the streets of Dublin to watch the passing of the funeral cortege, while tens of thousands witnessed the event as it was broadcast live on the national television network. With full state honours, the coffins, draped in the Irish tricolour, were publicly paraded from the gaol to the Catholic Pro-Cathedral in central Dublin where a concelebrated requiem mass was held before the bodies were transported for burial to Glasnevin cemetery. A graveside oration, delivered by the Irish *Taoiseach* (Prime Minister), was accompanied by the sounding of three rounds of ammunition and the playing of the *Last Post* and national anthem. While some controversy surrounded the day's events, by and large the ceremony was deemed a fitting, dignified and noble occasion of reconciliation and remembrance. The men concerned were Irish Republican Army (IRA) Volunteers executed eighty years earlier, under British authority, at Mountjoy gaol during the War of Independence 1920–21.[2] Their bodies had been buried in the grounds of the prison and their re-interring at Glasnevin cemetery had been mooted over subsequent decades. The final symbolic recognition of their sacrifice through the performance of a state funeral on a rainy autumnal day in 2001 reinforces the significance of the dead in the arousal of the collective and personal memories of the living.

In the Taoiseach's oration he claimed that: 'The big powers had said that it was for the small nations that the First World War was fought. The people of Ireland were determined that the principle of national self-determination must also be extended to the Irish nation.'[3] The lexical juxtaposition of the First World War

---

[1] There were actually ten men's bodies exhumed but Patrick Maher, at the request of his family, was re-interred in a cemetery in his home county of Limerick.

[2] The men executed were Kevin Barry, Thomas Whelan, Patrick Moran, Patrick Doyle, Bernard Ryan, Frank Flood, Thomas Bryan, Thomas Traynor, Edmund Foley and Patrick Maher.

[3] Bertie Ahern's (2001) graveside oration at Glasnevin cemetery, Sunday 14 October 2001. The text of the speech was published in full in the *Irish Times*, 15 October 2001.

with the question of Irish independence reminds us of the real proximity of the global conflict that was the Great War and the local conflict that was the Irish independence movement. The overlapping of these powerful political moments would be crucial for the development of a memorial landscape in Ireland to those who died in the Great War. Where the dead are concerned Verdery reminds us that 'Remains are concrete, yet protean,'[4] and it is precisely the protean nature of the rituals of remembrance dedicated to Irish men and women killed in the First World War that is the central concern of this study. This book situates efforts to publicly remember those who sacrificed their lives in the war within the context of a set of competing narratives of cultural identity evident in Ireland in the years preceding and following the war. This context acted both as a theatrical stage in which remembrance took place and a temporal stage in which rituals of public commemoration would be marked, rehearsed and repeated in the decades following the war.

Time, memory and representation

The central preoccupation of Al Pacino's late twentieth-century documentary movie *Looking for Richard* is making sense of a play written four centuries ago about an English king who reigned for two years. As an exercise in translation, Pacino's treatment of the play brings into sharp relief the challenges and possibilities presented by attempting to re-enact, re-stage, re-interpret and re-memorise an historical drama. The interpretation and performance of the play by an American cast, the location of the play in New York city and the conversations held between the cast, Shakespearean specialists, construction workers, high-school students and taxi drivers all underpin the questions that the movie raises about how the meaning of past events can be conveyed to contemporary audiences. The adverb of present time – Now – which dramatically introduces the opening speech of the play, delivered by Gloster, immediately unfetters the temporal chain of sequence usually deployed to evoke time's past and past times. To remember the past is to remember it now and each rehearsal of *Richard III* arises from the perspective of 'Now', and Pacino's search for meaning is one moment in that quest for meaning. From discussions of iambic pentameters, the internecine intrigue of the English court, the psycho-political and sexual motivations of the characters, the costuming of the actors, the War of the Roses, Pacino's documentary film makes transparent both the process of interpretation and the interpretation itself as it is represented by this particular cast. In so doing it makes visible the complex relationship between the context and text in any rendition of the past.

The translation of meaning across space and time is central both to the rituals of everyday life and to the exceptional moments of remembrance associated with

[4] K. Verdery, *The political lives of dead bodies: reburial and postsocialist change* (New York, 1999), 28.

birth, death and other key events in personal and collective histories. Memory as re-collection, re-membering and re-presentation is crucial in the mapping of historical moments and in the articulation of identity. As Jonathan Boyarin has put it 'memory is neither something pre-existent and dormant in the past nor a projection from the present, but a potential for creative collaboration between present consciousness and the experience or expression of the past'.[5]

Maurice Halbwachs' work *On Collective Memory* was the first critical attempt to give some sort of definition to the idea of social memory. For Halbwachs, collective or social memory was rooted in his belief that common memories of the past among a social group, tied by kinship, class or religion, link individuals in the group with a common shared identity when the memories are invoked. Social memory is a way in which a social group can maintain its collective identity over time and it is through the social group that individuals recall these memories.[6] But, as Withers has commented, this analysis itself is 'rooted in that concern for continuities evident in the *longue durée* tradition of French *Annaliste* historiography and in acceptance of a rather uncritical, "superorganic" notion of culture'.[7] While Halbwachs is right to socialise the concept of memory his analysis fails to historicise memory and embrace the notion that the very concept of the 'social' may itself have a history and indeed a geography.

Conventionally the 'art of memory' since Romanticism has been ideologically separated from history in Western historiographical traditions where memory is subjective, selective and uncritical while history is objective, scientific and subject to empirical scrutiny.[8] With the demise of peasant societies, Nora suggests that true memory 'which has taken refuge in gestures and habits, in skills passed down by unspoken traditions, in the body's inherent self-knowledge, in unstudied reflexes and ingrained memories'[9] has been replaced by modern memory which is self-conscious, historical and archival. More recent work on social memory has emphasised the discursive role of memory in the articulation of an identity politics and in particular the role of elite and dominant memory, mobilised by the powerful, to pursue specific political objectives.[10] The distinction between 'authentic' and

---

[5] J. Boyarin, *Remapping memory: the politics of timespace* (London, 1994), 22.

[6] M. Halbwachs, *On collective memory*, ed. and trans. L. Coser (Chicago, 1992). It was originally published in French as *La mémoire collective* (Paris, 1950).

[7] C. Withers, 'Place, memory, monument: memorializing the past in contemporary Highland Scotland', *Ecumene*, 3 (1996), 326.

[8] F. Yates, *The art of memory* (London, 1978).

[9] P. Nora, 'Between memory and history: les lieux de mémoire', *Representations*, 26 (1989), 13.

[10] There is a vast literature covering this theme but included as some of the most important are P. Hutton, *History as an art of memory* (Burlington, VT, 1993); J. Le Goff, *History and memory*, trans. S. Rendall and E. Clamen (New York, 1992); D. Krell, *Of memory, reminiscence and writing* (Bloomington, 1990); G. Lipsitz, *Time passages: collective memory and American popular culture* (Minneapolis, 1990); D. Middleton and D. Edwards, eds., *Collective remembering* (London, 1990). P. Nora, ed., *Realms of memory: Vol. 11: Traditions* (Chichester, 1997).

modern memory is particularly persuasive when connected with a style of politics associated with the rise of the national state. The development of extra-local memories is intrinsic to the mobilisation of an 'imagined community' of nationhood,[11] and new memories necessitate the collective amnesia or forgetting of older ones.[12] In particular, where elites are concerned Connerton suggests that 'it is now abundantly clear that in the modern period national elites have invented rituals that claim continuity with an appropriate historic past, organising ceremonies/parades and mass gatherings, and constructing new ritual spaces'.[13] In a fascinating study of the emergence of nationalist politics in Germany, Mossé investigates how the 'new politics' 'attempted to draw the people into active participation in the national mystique through rites and festivals, myths and symbols which gave concrete expression to the general will'.[14] Resisting analyses which focus primarily on the political and economic transformations which precipitated the evolution of the nation-state, Mossé's study shifts the historical emphasis towards the cultivation of a collective memory by focusing on the aesthetics and symbolism central to German nationalism. He claims: 'it [nationalism] represented itself to many, perhaps most people, through a highly stylised politics, and in this way managed to form them into a movement'.[15] As such, the role of re-membering the past – the putting together of its constituent parts into a single, coherent narrative – has been profoundly significant for the emergence of a popular nationalist identity. The deployment of the body as an analogy of the nation-state, a genealogy of people with common origins, co-exists with a claim that the state acts as a guarantor of individual rights and freedoms that transcend historical time and the constraints of the past. Paradoxically, then, in the context of national identity, social memory as mediated through political elites both legitimates and simultaneously denies the significance of remembrance of things past.

While, at its most basic level, memory can be said to operate at the scale of the individual brain and thus avoid a concept of memory that suggests it has a superorganic quality, it is also necessarily the case that memories are shared, exchanged and transformed among groups of individuals. In this sense there are collective memories which arise from the inter-subjective practices of signification that are not fixed but are re-created through a set of rules of discourse that are periodically contestable.[16] This can be seen, for instance, in the collective memory of the American Civil War as expressed through the erecting of public statues. Rather than reflecting the serious division between pro- and anti-slavery lobbies in the United States, they were gradually perceived 'as part of a healthy process of sectional reconciliation – a process that everyone knew but no one said was for and

---

[11] B. Anderson, *Imagined communities: on the origins and spread of nationalism* (London, 1989).

[12] On the idea of the invention of national traditions, see the seminal work E. Hobsbawm and T. Ranger, eds., *The invention of tradition* (Cambridge, 1983).

[13] P. Connerton, *How societies remember* (Cambridge, 1989), 51.

[14] G. Mossé, *The nationalization of the masses* (New York, 1975), 2.   [15] *Ibid.*, 214.

[16] J. Butler, *Gender trouble: feminism and the subversion of identity* (London, 1990).

between whites'.[17] The context of signification in this case was the reconciliation of northern and southern whites in the rules of a discourse, which denied black memory and thus blurred the South's defence of slavery. This visual interpretation of the Civil War, however, did not exist completely uncontested and there were three statues erected to represent blacks. Two of these monuments displayed a single black soldier amongst a group of combatants. The third – the Shaw memorial – erected in Boston in 1897 and designed by the sculptor Augustus Saint-Gaudens, was of the commander Robert Gould Shaw surrounded by his regiment of black troops. This facilitated 'opposing readings of its commemorative intent'[18] and underlines the periodic capacity for memories to be contested in the public sphere.

There is a considerable literature emphasising the politics of memory, especially where dominant groups in society are concerned, *vis-à-vis* their shaping of interpretations of the past. Yet it is increasingly clear that the social process involved in memorialisation is hotly contested with respect not only to form and structure but also to the meaning attached to the representation. Popular memory can be a vehicle through which dominant, official renditions of the past can be resisted by mobilising groups towards social action but also through the maintenance of an oppositional group identity embedded in subaltern memories.[19] The deployment of local and oral histories in the formation of group identities can be a powerful antidote to both state and academic narratives of the past, especially where marginalised groups are concerned.[20] The controversies surrounding the remembering of the Holocaust through the conversion of death camps into 'memorial' camps to the genocide of the Second World War is a case in point. In Auschwitz, for instance, the competing aspirations of Polish nationalists, communists, Catholics and Jews to control the representation of the Holocaust there has influenced the physical structure of the site and the meaning attached to it by these various groups.[21] In this sense, rather than treating memory as the manipulative action of the powerful to narrate the past to suit their particular interests, a fuller account might follow Samuel who suggests that one 'might think of the invention of tradition as a process rather than an event, and memory, even in its silences, as something which people made for themselves'.[22] The capacity which people have to formulate and represent their

---

[17] K. Savage, 'The politics of memory: black emancipation and the Civil War monument', in R. Gillis, ed., *Commemorations: the politics of national identity* (Princeton, 1994), 132.

[18] *Ibid.*, 136.

[19] R. Johnson, G. McLennan, B. Schwarz and D. Sutton, eds., *Making histories: studies in history-writing and politics* (London, 1982).

[20] See, for instance, K. Armstrong and H. Benyon, eds., *Hello are you working?! Memories of the thirties in the north east of England* (Durham, 1977).

[21] See A. Charlesworth, 'Contesting places of memory: the case of Auschwitz', *Environment and Planning D: Society and Space*, 12, (1994), 579–93; J. E. Young, *The texture of memory: holocaust memorials and meaning* (London, 1993); H. Langbein, 'The controversy over the convent at Auschwitz', in C. Rittner and J. K. Roth, eds., *Memory offended: the Auschwitz convent controversy* (New York, 1991), 95–8.

[22] R. Samuel, *Theatres of memory*, vol. I (London, 1994), 17.

own memories, however, is regularly constrained by the discursive field in which they operate and literally the space in which their pronouncements, both figurative and literal, are made. As Sherman reminds us, 'commemoration is also cultural: it inscribes or reinscribes a set of symbolic codes, ordering discourses, and master narratives that recent events, perhaps the very ones commemorated, have disrupted, newly established, or challenged'.[23] If memory is conceived as a recollection and representation of times past, it is equally a recollection of spaces past where the imaginative geography of previous events is in constant dialogue with the current metaphorical and literal spatial setting of the memory-makers.

Space, memory and representation

The role of space in the art and the act of memory has a long genealogy in European thought. In the ancient and medieval worlds memory was treated as a visual rather than a verbal activity, one which focused on images more than words. The immense dialectal variation and low levels of literacy perhaps account for the primacy of the visual image over other types of representation. Visual images like the stained glass window and other religious icons came to embed a sacred narrative in the minds of their viewers. They became mnemonic devices in religious teaching where sacred places became symbolically connected to particular ideal qualities. Networks of shrines, pilgrimage routes and grottoes, sited for commemorative worship, formed a sacred geography where the revelations of a Christian God could be remembered, spatially situated and adored.[24] A mapping of the narrative of Christianity through a predominantly visual landscape formed the basis of memory work through the Middle Ages.[25]

While during the Renaissance and Enlightenment the conception of memory work altered scale (to the astral) and focus (towards the scientific rather than the religious), and was expressed at times architecturally by viewing the world from a height,[26] it was during the period of Romanticism that a more introspective, personal and localised view of memory came into focus. Memory in this guise came to be seen as the recovery of things lost to the past, the innocence of childhood and childhood spaces, for instance, and this divorced memory work from any scientific endeavour to make sense of the world or the past. It transformed the role of memory to the scale of the individual and perhaps created the preconditions for divorcing history from memory and separating intellectually the objective spatial narratives of history from the subjective experience of memory places. But as Samuel persuasively argues, 'far from being merely a passive receptacle or storage system, an image bank of the past, [memory] is rather an active, shaping force; that it is dynamic – what it contrives symptomatically to forget is as important as

---

[23] D. Sherman, *The construction of memory in interwar France* (London, 1999), 7.

[24] M. Carruthers, *The book of memory: a study of memory in medieval culture* (Cambridge, 1990).

[25] B. Kedar and R. Werblowsky, eds., *Sacred space: shrine, city, land* (New York, 1998).

[26] Yates, *The art of memory*.

what it remembers – and that it is dialectically related to historical thought, rather than being some kind of negative other to it'.[27]

By treating memory as a dialectic of history, in constant dialogue with the past, we begin to see how the dualistic thinking underwriting the division of history and memory becomes more problematic. This is particularly the case in relation to the spatiality of history and memory. The gradual transformation of a sacred geography of religious devotion to a secularised sacred geography connected with identity in the modern period destabilises the rigid lines of demarcation drawn between objective/subjective narration; emotional/abstract sources of evidence; local/universal ways of knowing. Treating memory as a legitimate form of historical understanding has opened new avenues of research where subjective renderings of the past become embedded in the processes of interpretation and not as a counterpoint to objective facts. Nation-building exercises, colonial expansion in the non-European world, regional, ethnic and class identity formation, all embrace an imaginative and material geography, made sacred in the spaces of remembrance and continuously remade, contested, revised and transmuted as fresh layers of meaning attend to the spaces. Geographers, historians, anthropologists and cultural theorists are increasingly paying attention to the processes involved in the constitution and routing of memory spaces, and especially to the symbolic resonances of such spaces to the formation, adaptation and contestation of popular belief systems.

In particular, studies have focused on the role of commemorative spaces and memory making in the articulation of national identity. In the context of the United States, the intersections between vernacular and official cultural expressions have been demonstrated to create a series of commemorative sites and rituals which attempt to combine some of the divergent sources of memory (e.g. local, ethnic, gender) with nationalising ones. The vocabulary of patriotism is particularly important 'because it has the capacity to mediate both vernacular loyalties to local and familiar places and official loyalties to national and imagined structures'.[28] Similarly, because of the divergent allegiances generated by specific sites of memory, they operate multivocally and are read in divergent and at times contradictory ways. The commemoration of the American Civil War points to the underlying fissures evoked by remembrance of a divisive episode in a state's history. The spatiality of memory is not only mirrored in the physical distribution of commemorative sites but also in the interpretative apparatus embedded in them. For instance, the commemorative statue to General Lee in Richmond, Virginia focuses on his role as an American hero who fought out of loyalty to his home state and obscures the larger political and racial politics which undergirded the war.[29] The

[27] Samuel, *Theatres of memory*, vol. I, x.

[28] J. Bodnar, *Remaking America: public memory, commemoration and patriotism in the twentieth century* (Princeton, 1992), 14–15.

[29] For a discussion of Civil War monuments see S. Davis, 'Empty eyes, marble hand: the Confederate monument and the South', *Journal of Popular Culture*, 16 (1982), 2–21; G. M. Foster, *Ghosts of the Confederacy: defeat, the lost cause, and the emergence of the new South* (Oxford, 1987); H. E. Gulley,

equestrian statue on Monument Avenue was part of a larger speculative real-estate venture where an expensive residential subdivision of property was laid out along the long avenue. Linking business, art and memorywork, the 'legitimation of Lee in national memory helped erase his status as traitor, as "other", leaving otherness to reside in the emancipated slaves and their descendants, who could not possibly accept Lee as their hero'.[30] The controversy surrounding the siting, design and iconographic effect of the Vietnam Veterans' Memorial in Washington DC is also an exemplary case. The public's ambiguous response to America's role in the war was further highlighted in attempts to commemorate the event. The heated debate underpinning the choice of design and designer, combined with the siting of the memorial along the Mall – a thoroughfare of national remembrance – reveals the regional, ethnic, social and gender tensions that this act of memorialisation brought to the surface.[31]

Discussions of nation-building projects and the memory spaces associated with them have been analysed as a form of mythology – a system of story-telling in which that which is historical, cultural and situated appears natural, innocent and outside of the contingencies of politics and intentionality. Drawing from semiology and linguistics such work claims that 'the apparent innocence of landscapes is shown to have profound ideological implications . . . and surreptitiously justif[ies] the dominant values of an historical period'.[32] Geographers have extensively explored the promotion of specific landscape images as embodiments of national identity.[33] Historians have paid attention to the evolution of particular festivals, rituals,

'Women and the lost cause: preserving Confederate identity in the American Deep South', *Journal of Historical Geography*, 19 (1993), 125–41; J. J. Winberry, 'Symbols in the landscape: the Confederate memorial', *Pioneer America Society Transaction*, 5 (1982), 9–15; J. J. Winberry, ' "Lest we forget": the Confederate monument and the southern townscape', *Southeastern Geographer*, 23 (1983), 107–21.

[30] Savage, *The politics of memory*, 134. The latest episode in the memorialising of Monument Avenue is found in J. Leib, 'Separate times, shared spaces: Arthur Ashe, Monument Avenue and the politics of Richmond, Virginia's symbolic landscape', *Cultural Geographies*, 9 (2002), 286–312.

[31] For a full discussion of the controversy see R. Wagner-Pacifini and B. Schwartz, 'The Vietnam Veterans' Memorial: commemorating a difficult past', *American Journal of Sociology*, 97 (1991), 376–420; M. Sturken, 'The wall, the screen and the image: the Vietnam Veterans Memorial', *Representations*, 35 (1991), 118–42.

[32] J. S. Duncan and N. G. Duncan, 'Ideology and bliss: Roland Barthes and the secret histories of landscape', in T. Barnes and J. S. Duncan, *Writing worlds: discourse, text and metaphor in the representation of landscape* (London, 1992), 18.

[33] See, for instance, the special issue of the *Journal of Historical Geography*, 'Creation of myth: invention of tradition in America', ed. J. L. Allen, 18 (1992), 1–138; M. Azaryahu, 'From remains to relics: authentic monuments in the Israeli landscape', *History and Memory*, 5 (1993), 82–103; M. Heffernan, 'For ever England: the Western Front and the politics of remembrance in Britain', *Ecumene*, 2 (1995), 293–324; Withers, 'Place, memory, monument', 325–44; R. Peet, 'A sign taken from history: Daniel Shay's memorial in Petersham, Massachusetts, *Annals of the Association of American Geographers*, 86 (1996), 21–43; M. Auster, 'Monument in a landscape: the question of "meaning" ', *Australian Geographer*, 28 (1997), 219–27; M. S. Morris, 'Gardens "Forever England":

public holidays and so on in the evolution of the 'myth' of nationhood.[34] Others have explored the social relations underpinning a particular landscape. Schorske's exploration of the nineteenth-century redesign of the Ringstrasse in Vienna as a 'visual expression of the values of a social class'[35] meshes a discussion of the economic and political with the aesthetic in the reconceptualisation of the urban form. While Harvey's analysis of the Basilica of Sacré-Coeur in Paris refashions our understanding of that space by emphasising its connections with the tumultuous class politics of that city in the nineteenth century, it also reminds us that what the basilica stands for is not readily clear from the representation itself.[36] The materiality of a particular site of memory sometimes masks the material social relations undergirding its production by focusing the eye on its aesthetic representation independent of the sometimes less visible ideas (social, economic, cultural power relations) underlying the representation. It is often then in the realm of ideas, however contested and contradictory, that the meaning of memory spaces is embedded. What idea or set of ideas are stimulated by memories made material in the landscape?

The emphasis on visual interpretations of the memory landscapes that undergirded medieval sacred geographies continues to animate discussions of landscape interpretation today. The treatment of a landscape as a text which is read, and actively reconstituted in the act of reading as the 'context of any text is other texts',[37] including conventional written texts as well as political and economic institutions, reinscribes the visual as the central action of interpretation.[38] While offering a more nuanced understanding of the act of reading any landscape and the possibility of decoding the messages within any space, the text metaphor may overemphasise the power to subvert the meaning of landscape through its reading, without necessarily providing a space in which to change the landscape itself. Hegemonic and subaltern readings, in other words, may take precedence over hegemonic and subaltern productions.[39] In the context of the First World War, for instance, the

---

landscape, identity and the First World War cemeteries on the Western Front', *Ecumene*, 4 (1997), 410–34; D. Atkinson and D. Cosgrove, 'Urban rhetoric and embodied identities: city, nation and empire at the Vittorio Emanuele II monument in Rome, 1870–1945', *Annals of the Association of American Geographers*, 88 (1998), 28–49.

[34] Hobsbawm and Ranger, *The invention of tradition*. See also R. Porter, ed., *Myths of the English* (Cambridge, 1992).

[35] C. E. Schorske, *Fin-de-siècle Vienna: politics and culture* (London, 1979), 25.

[36] D. Harvey, 'Monument and myth', *Annals of the Association of American Geographers*, 69 (1979), 362–81.

[37] J. S. Duncan, *The city as text: the politics of landscape interpretation in the Kandyan Kingdom* (Cambridge, 1990).

[38] For a full discussion of the text metaphor see T. Barnes and J. Duncan, eds., *Writing worlds: discourse, text and metaphor in the representation of landscapes* (London, 1992); J. Duncan and N. Duncan, '(Re)reading the landscape', *Environment and Planning D: Society and Space*, 6 (1988), 117–26; J. Duncan and D. Ley, eds., *Place/culture/representation* (London, 1993).

[39] D. Mitchell, *Cultural geography: a critical introduction* (Oxford, 2000).

desire to forget, erase and bury the memory of the war among veterans may have run contrary to the desire to remember, erect and exhume the memory of the war among non-combatants. The focus on the metaphor of the text also tends to underestimate the aural dimension of texts where, in the past, reading was a spoken activity. Reading texts aloud where the sounds, rhythms and syntax of the words are collectively absorbed directs attention to the social nature of interpretation which embraces senses other than the purely visual. Treating the landscape as a theatre or stage broadens the imaginative scope of interpretation by suggesting that life gets played out as social action and social practice as much as it does by the reading implied by the text metaphor. As Cosgrove argues, 'landscapes provide a stage for human action, and, like a theatre set, their own part in the drama varies from that of an entirely discreet unobserved presence to playing a highly visible role in the performance'.[40] This notion of landscape as theatre could be further extended, not solely as the backdrop in which the action takes place but as actively constituting the action. The stage acts more than as the context for the performance; it is the performance itself.

The idea of life as drama played out through spectacle is particularly helpful when considering the memory of war. Where spectacle is concerned, 'It could take on the sense of a mirror through which truth which cannot be stated directly may be seen reflected and perhaps distorted.'[41] To make sense of the drama of intense physical conflict and the human losses attendant on it requires both dramatic and silent modes of remembrance. That romantic notions of memory seemed inadequate to deal with the losses of the First World War is evidenced by the fact that enormous collective and individual efforts were made to articulate that sense of loss through public performance. From literary texts that had widespread circulation to the massive war cemeteries created in France and elsewhere, the very technology of modernity that facilitated such a massive loss of life also facilitated acts of mass commemoration.[42] Nonetheless, to represent such events was to try to make sense of them while simultaneously engaging in the very crisis of representation that the pain of war engendered. This book is precisely concerned with the variety of ways in which the First World War was represented – the silent and noisy spaces of remembrance which constituted the Irish context.

---

[40] D. Cosgrove, *The Palladian landscape: geographical change and its cultural representations in sixteenth century Italy* (University Park, PA, 1993), 1.

[41] S. Daniels and D. Cosgrove, 'Spectacle and text: landscape metaphors in cultural geography', in Duncan and Ley, *Place/culture/representation*, 58.

[42] For studies dealing with mass commemoration see for Britain A. Gaffney (1998), *Aftermath: remembering the Great War in Wales* (Cardiff, 1998); A. Gregory, *The silence of memory: Armistice Day 1919–1946* (Oxford, 1994); A. King, *Memorials of the Great War in Britain* (Oxford, 1998); for Australia see K. S. Inglis, *Sacred places: war memorials in the Australian landscape* (Melbourne, 1998); for France see A. Prost, 'Monuments to the dead', in Nora, ed., *Realms of memory*, vol. II, 307–32; D. Sherman, *The construction of memory in interwar France* (London, 1999).

Remembering the First World War

While the First World War has generated a vast academic and popular literature, much of the discussion of the memory of it has been sparked by the thesis originating with Paul Fussell's book *The Great War and Modern Memory* (1975). Fussell claims that the conflict marked a watershed in European conceptions of war where the old certainties and formulaic languages of duty and heroism were replaced by ironic, negative and darker visions of the human spirit. Drawing primarily on literary sources, Fussell's book tracks the languages of ironic modernism that are found in the prose, novels and poetry of the war's literary soldiers.[43]

Others have followed this line of argument and have exemplified, in a variety of national contexts, how the direct experience of war by writers as combatant soldiers translated the war in a fashion far removed from the 'high diction' and patriotic rhetoric that informed the older generation of writers, generals and political leaders.[44] Critics of this position have pointed to the unrepresentative nature of Fussell's sources, that is, based on the evidence of white Anglo-American males with literary aspirations who served on the front lines.[45] Feminist historians have queried the thesis that the war proffered radical changes in value systems and they have highlighted the ambiguity of the gains enjoyed by women in the inter-war years.[46] Studies of women's experience during the war similarly reveal the challenge to feminine identity that the war both demanded and tried to restrict, and how this process was negotiated in complicated ways.[47] Drawing from more mundane literary sources than those influenced by modernist theses, recent scholars have suggested that conservatism and tradition persisted in the inter-war years and that in many ways the war represented continuity rather than radical discontinuity.[48] In a brilliant discussion of Canada's remembrance of the war, Vance powerfully elucidates how an official public memory and an unofficial private one were frequently intertwined in Canada's articulation of a social memory, and writes that 'Canadians were concerned first and foremost with utility: those four years had to have been of some use.'[49] They did this by emphasising the very tropes of duty, righteousness, sacrifice and redemption that modernists have depicted as spent forces.

---

[43] P. Fussell, *The Great War and modern memory* (Cambridge, 1975).

[44] M. Eksteins, *Rites of spring: the Great War and the birth of the modern age* (New York, 1989); S. Hynes, *A war imagined; the Great War and English literature* (London, 1991).

[45] L. Hanley, *Writing war: fiction, gender and memory* (Amherst, MA, 1991).

[46] M. Higonnet, J. Jenson, S. Michel and M. Weitz, eds., *Behind the lines: gender and two world wars* (London, 1987).

[47] S. Ouditt, *Fighting forces, writing women: identity and ideology in the First World War* (London, 1994).

[48] See R. M. Bracco, *Merchants of hope: British middlebrow writers and the First World War, 1919–39* (Oxford, 1993); D. Englander, 'Soldiering and identity: reflections on the Great War', *War in History*, 1 (1994), 300–18.

[49] J. F. Vance, *Death so noble: memory, meaning and the First World War* (Vancouver, 1997), 9.

The most trenchant critique of the modernist thesis is provided by Jay Winter in his fascinating analysis of sites of memory. While Winter does not seek to underestimate the significance of modernism to the early twentieth century more generally and to the war in particular, he is also convinced that the language and practices of tradition – religious motifs, romantic forms, classical designs – continued to find expression and value in the years following the conflict. His scepticism of a radical break thesis resides in the historiographical point that 'To array the past in such a way is to invite distortion by losing a sense of its messiness, its non-linearity, its vigorous and stubbornly visible incompatibilities.'[50] And he also contends that although the ironic and cynical representations of war could convey anger and despair at the huge loss of life, they could not have healing power. It is precisely the capacity of the language of tradition to provide a sense of solace for grieving families and friends that provided it with its popular impetus in the creation and maintenance of sites of memory dedicated to the war. Winter's concern is to highlight some of these across a variety of national contexts. It is perhaps the coexistence of traditional and modernist modes of representation – the desire to simultaneously remember and to forget – that marks war as a particular arena of memory that is laced with contradictions and disputes. That the public expression of grief was interspersed with the private and that the spaces normally used for public actions also became the spaces for very private mourning muddied the role of space in the articulation of private and public lives.

Geographers and others who have examined the creation of landscapes of memory for soldiers have highlighted just how many debates surrounded such acts of representation and how contested the images and practices of remembrance have been.[51] This book is concerned with examining the articulation of remembrance in a society itself in political and cultural turmoil during and immediately after the war. The narrative of war commemoration in Ireland was consistently in dialogue with the narratives attendant on the national question. The war did not represent in Ireland an opportunity for the divergent voices of Irish nationalism and unionism to unite. Unlike the suffragist movement in Britain, for instance, which rallied behind the war for its duration, in Ireland the war ironically became

---

[50] J. Winter, *Sites of memory, sites of mourning: the Great War in European cultural history* (Cambridge, 1995), 5.

[51] K. Till, 'Staging the past: landscape design, cultural identity and *Erinnerungspolitik* at Berlin's Neue Wache', *Ecumene*, 6 (1999), 251–83; J. Bell, 'Redefining national identity in Uzbekistan: symbolic tensions in Tashkent's official public landscape', *Ecumene*, 6 (1999), 183–213; H. Leitner and P. Kang, 'Contested urban landscapes of nationalism: the case of Taipei', *Ecumene*, 6 (1999), 172–92; B. Osborne, 'The iconography of nationhood in Canadian art', in D. Cosgrove and S. Daniels, eds., *The iconography of landscape* (Cambridge, 1988), 162–78; B. Osborne, 'Figuring space, marking time: contested identities in Canada', *International Journal of Heritage Studies*, 2 (1996), 23–40; B. Osborne, 'Warscapes, landscapes, inscapes: France, war, and Canadian national identity', in I. Black and R. Butlin, eds., *Place, culture and identity* (Quebec, 2001), 311–33; S. Cooke, 'Negotiating memory and identity: the Hyde Park Holocaust Memorial, London', *Journal of Historical Geography*, 26 (2000), 449–65.

part of the vehicle through which the disparate voices of identity politics found expression.[52]

From the recruitment campaigns in the early years of the war to the commemorative rituals following the armistice, Ireland's role in the war was consistently interpreted through the lens of the conflicting tropes of identity on the island. Individual grief could not be separated from the larger canvas in which memory was mobilised. The neat binaries of victor and vanquished, enemy and friend, Christian and heathen, public and private, individual and collective collapsed during the war and in the years following it. And this collapse found expression in the very spatiality of memory. The sites in which collective memory could be rooted became in themselves the sight-lines through which the conflict would be viewed. The divisions in the national imaginary, present before the war, were heightened and accentuated as the memory of the war was materialised in rituals, memorials and literary texts in the post-war period. And these divisions did not operate solely at the scale of the social group but they also encompassed a schizophrenic attitude of mind for the individual. That there was a rebellion on Irish soil during the war, a war of independence in the years immediately after the armistice, partition of the island in 1921, and a subsequent civil war in the Irish Free State, all testify to the complex local circumstances which underpinned efforts to create a landscape of remembrance.

Yet despite these conflicting narratives of identity there were public acts of commemoration and it is unravelling the debates surrounding these that is the principal concern of this book. The following five chapters will be concerned with the *stages* of memory both in the sense of the theatrical metaphor where the spectacle of life and the work of memory is enacted, but also in the temporal sense of transmutation of meaning over time. There were stages of reaction to the war, from the innocent optimism of new recruits volunteering in 1914, followed by periods of pessimism and depression surrounding long phases of stalemate, to the post-war grieving of veterans and bereaved families. In Ireland the war represented opportunity and postponement; quiet support and loud dissent; active participation and passive observation; victory and defeat. In what was to become the Irish Republic, the hyper-spectacle that animated the memory work of many other countries – the proliferation of monument, memorial and ceremony, the literature, the annual parade, the historiography – did not take hold to the same extent. It is precisely this ambiguity between remembrance and forgetting that is the subject of this book.

The following chapters each deal with a particular aspect of memory making and each attempts to identify how the idea and act of remembrance in an Irish context was articulated in complex ways. This is not to engage in an exercise of national exceptionalism. It is to make the case that a geography of remembrance

---

[52] K. Jeffery, *Ireland and the Great War* (Cambridge, 2000); I. McBride, ed., *History and memory in modern Ireland* (Cambridge, 2001).

is important even within the universalising languages of bitter irony or painful sorrow. I have selected a number of critical moments in the making of popular memory and in that sense this book does not represent a strict chronology of remembrance nor is it exhaustive. Instead it seeks to narrate the commemoration of the war through a selection of key episodes. These largely took place in the first two decades after the war when much of the memory work was established. As a contextual framework, however, Chapter 2 provides the backdrop for the war in Ireland. Situating the war in its political and cultural context, this chapter examines how an army was recruited on the island and how persuasive images were circulated to entice Irish men into the army in the shadow of the highly variable levels of loyalty to the union of Britain and Ireland. Chapters 3, 4 and 5 each take a strategic episode of remembrance activity – the parade, the memorial, the literary text – and explicates the debates and acts of memory work that were performed in the years following the war. Each of these is placed in the context of the changing political geography of the island with particular focus on the narratives of commemoration in what would become the Irish Republic. This book will mobilise some of the divergent approaches to spatialising memory in the north of Ireland (pre- and post-partition) as a counterpoint to the patterns which emerged in the south. Rather than offering a comprehensive account of the politics of memory in Northern Ireland, these comparisons will serve to highlight the significance of geography to the construction of memory on the island. Chapter 6 juxtaposes Ireland's remembrance of the war with its memorialisation of the 1916 Easter Rebellion. Due to the significance attached to the rebellion in historiographical and popular terms, an analysis of its role in the mapping of national memory will serve to spotlight the different debates attendant on its remembrance, particularly as celebrations reached their apotheosis during the fiftieth anniversary. Overall, commemorating one war in the wake of a rebellion, a guerrilla struggle and subsequent civil war, and in the shadow of a newly emerging state, all played upon the manner in which the First World War could be forgetfully remembered in Ireland.

2

A call to arms: recruitment poster and propaganda

If, as some authors argue, the First World War marked a 'satire of circumstance'[1] for the young men and women of Europe in the second decade of the twentieth century, for Irish people the events of 1914–18 marked no less a panoply of contradictions. Characterised as the first modern war where technology and communications enhanced, on scales heretofore never witnessed, the capacity to obliterate life with extreme regularity and ferociousness,[2] the recruitment needs of all sides in the conflict implicated sections of the population which, until then, were immune from military experience and modern warfare.

Enacted, to a great extent, by a volunteer army, recruited and trained 'for the duration' and whose commitment to military life was to extend no longer than the conflict, the war necessitated that the state undertake a massive drive to enter into the hearts and minds of young men and women whom it sought to recruit. While to 'fight for one's country' or one's empire was not in itself a new phenomenon (indeed the heroic soldier of the literary consciousness had entered the imagination of the young long before the war),[3] the call to arms nevertheless was a structured, planned activity which, through a variety of means, sought to tap into a suite of cultural and political prejudices of the day.[4] Despite the common references found in recruitment propaganda of all participating states, there are, equally, strategies in this literature which reveal the different recruitment policies observed by individual combatant states. France, Britain and the United States, although allied in the field of battle, exercised individual discretion where eliciting popular support for the war was concerned.

Thus, although a common cause could be identified, the diverse cultural milieux in which the war was enacted necessitated individual states adopting strategies

---

[1] The expression comes from Fussell, *The Great War and modern memory*.
[2] For a full analysis of casualty figures see J. M. Winter, *The Great War and the British people* (London, 1985).
[3] Hynes, *A war imagined*.
[4] See M. Hardie and A. K. Sabin, *War posters* (London, 1920); M. Rickards, *Posters of the First World War* (London, 1968); B. Hillier, *Posters* (London, 1969).

which were in sympathy with the local, regional, national and imperial loyalties of their constituent parts. Recruitment was, thus, a spatial activity, operating at a variety of scales. The proximity of Ireland to the British military planning centre, its political union with the Crown, and its participation in previous international conflicts (for instance, the Boer War) made it no less a ripe terrain for the recruitment of the New Armies in 1914. The conflict in political ideology in Ireland, unlike other parts of the Union, generated diverse responses to recruitment and contrasting attitudes to Irish participation in the war. While the social memory of war was produced largely in the years following the end of the conflict, in terms of the imaginary of war the enlistment of soldiers through the recruitment strategies of the states involved provides insights into the larger discourse of war. The suite of visual images, texts and speeches that circulated widely throughout the war years provide a foundation for the ways the war would subsequently be remembered. This chapter seeks to outline and provide an interpretation of the dominant motifs of war that were embedded in the Irish recruitment campaigns and to highlight how the broader discourse of a 'just war' was domesticated to meet local needs specific to the Irish context. In so doing this chapter seeks to elucidate how the war got scripted through the recruitment efforts and how, on the one hand, this scripting sought to accommodate political differences in Ireland, while ironically, at the same time, providing the very bases for opposition and dissent to the war in certain circles.

Political and cultural background

There were, of course, existing Irish regiments and a standing army in Ireland, from which initial mobilisation took place.[5] These comprised men who for economic and other reasons had chosen the military life and who formed part of the British Expeditionary Force dispatched to France and Belgium in August 1914. Each of the eight Irish regiments of the regular army had two battalions and the Irish Guards had one. There were four Irish cavalry regiments but the static pattern, which would characterise the conflict, especially on the Western Front, reduced the role of the cavalry and many troops were reassigned to support the infantry. The geographical hinterland of each of the regiments, although never recruiting solely from within their catchment area, acted as a guide to the geographical basis of recruitment.[6] They were as follows:

Royal Irish Regiment – South East Ireland
Royal Inniskilling Fusiliers – Donegal, Derry and mid-Ulster
Royal Irish Rifles – Belfast, Antrim and Down
Royal Irish Fusiliers – Armagh, Monaghan and Cavan

---

[5] H. E. Harris, *The Irish regiments in the First World War* (Dublin, 1968).

[6] M. Dungan, *Distant drums: Irish soldiers in foreign armies* (Belfast, 1993); T. Denman, 'Irish politics and the British army list: the formation of the Irish Guards in 1900', *The Irish Sword*, 19 (1995), 77, 171–86.

Connaught Rangers – Connaught
Leinster Regiment – Leinster
Royal Munster Fusiliers – Munster
Royal Dublin Fusiliers – Dublin and hinterland

While some of these regiments have a very long history (for instance, the Irish Guards was raised by Charles II in 1662, later disbanded and redistributed to other regiments), others were founded during the Williamite and Napoleonic wars (for instance, the Connaught Rangers).[7] For the first time during the South African War (1899–1902) Queen Victoria rewarded Irish soldiers for their contribution to the war effort through issuing shamrocks to each of the Irish regiments on St Patrick's day. She also established a new regiment of Foot Guards (later labelled the Irish Guards).

All the Irish regiments were represented in the British Expeditionary Force sent to the European continent during August and September 1914. While several battalions were either in barracks around Ireland or on training exercises, all responded to the call to mobilise. The precise motivation for men to enlist in the regular army is unclear, although unemployment or underemployment may have been important motivating factors.[8] A corporal from Cork with the Royal Irish Rifles observed that in his battalion, there

was an ex-divinity student with literary tastes; a national school teacher; a man who had absconded from a colonial bank; a few decent sons of farmers. The remainder of us in our Irish regiment were either scallawags or very minor adventurers.[9]

It may be impossible to trace the motives behind regular soldiers' commitment to the war, but the volunteer army forms a much more fertile ground for understanding the meaning of the war. The cultural mediation of the war through the state's recruitment agencies in tandem with the political context in Ireland during this period provides the contextual backdrop for enlistment.

[7] H. F. N. Jourdain, *History of the Connaught Rangers*, 3 vols. (London, 1925–8); A. E. C. Bredin, *History of the Irish soldier* (Belfast, 1987); M. Cunliffe, *The Royal Irish Fusiliers, 1793–1950* (Oxford, 1971); S. P. Kerr, *What the Irish regiments have done* (London, 1916).
[8] While there are no precise unemployment figures for Ireland in the years leading up to the war, there is continued evidence of large-scale emigration. The population declined from 6,552,385 in 1851 to 4,390,219 in 1911 and figures for provincial emigration during this period reveal the highest levels of out-migration occur in Ulster and Munster. See G. Doherty, 'Post-famine emigration' in S. Duffy, ed., *Atlas of Irish History* (Dublin, 1997), 102–3. Similarly, case studies reveal evidence of localised poverty, underemployment or insecure employment. In Dublin, for instance, Daly claims 'Two periods stand out as marked by serious unemployment: the early 1880s and the years from 1904–12', 107. In a Dublin context she remarks that: 'The basic labouring wage of approximately £1 per week was inadequate to support any family with children in the absence of supplementary earnings' (112). This was reflected in the high levels of tenement occupancy, with more than 25 per cent of families living in one-room flats in 1911 of more than four persons. For a fuller account see M. Daly, *Dublin: The deposed capital. A social and economic history 1860–1914* (Cork, 1984).
[9] M. Dungan (1995), *Irish voices from the Great War* (Dublin, 1995), 17.

Full steam ahead, John Redmond said
That everything was well chum;
Home Rule will come when we are dead
And buried out in Belgium[10]

While there was a regular army stationed in Ireland in 1914, it was the four other 'unofficial' armies whose roles would be central in fashioning Ireland's response to the war effort. To understand the existence of these competing military units, we need to retrace our steps briefly into the political and cultural climate in Ireland in the years leading up to the war.

The death of Parnell in 1891 and the split among constitutional nationalists led to a decade or more of division and political in-fighting. This was matched by a period of cultural renewal expressed through the literary movement of the Celtic Revival and the establishment of the Gaelic League to stimulate interest in the Irish language.[11] The complex interrelationships between the cultural and political movements are evidenced by the fact that the leadership and membership of each sometimes overlapped.[12] The simple distinctions between the political and the cultural arena cannot be sustained when examination is made of the policies and practices of each group.[13] Eoin MacNeill, for instance, an Ulsterman and academic by profession, was a co-founder of the Gaelic League as well as becoming leader of the Irish Volunteers and later a minister in the Cabinet of the first government in the Irish Free State.[14]

With the parliamentary defeat of the Home Rule Bills of 1886 and 1893, Ulster unionists had established their opposition to any form of devolved government in Ireland. The 1910 general election, which brought the Liberal Party to power, but dependent on the support of John Redmond's reunited Irish Parliamentary Party, stimulated renewed attempts to reintroduce a Home Rule bill. The introduction of the Parliament Act 1911, which replaced the House of Lords' right to veto bills with one of delaying parliamentary legislation, heightened anxieties among the unionist population in Ireland.[15] In April 1912 when a new Home Rule Bill was introduced in parliament it immediately aroused opposition in Conservative and Unionist circles. Although the seeds of unionist opposition pre-date the bill, the years 1912–14 would prove decisive in consolidating division on the island. Drawing support 'for Ulster Unionism in the House of Lords, the army and the

---

[10] *The Worker's Republic*, 6 November 1915.

[11] S. O Tuama, ed., *The Gaelic League idea* (Dublin, 1972); B. O Cuív, ed., *A view of the Irish language* (Dublin, 1969).

[12] J. Hutchinson, *The dynamics of cultural nationalism: the Gaelic Revival and the creation of the Irish nation state* (London, 1987).

[13] *Ibid.*

[14] F. X. Martin and F. J. Byrne, eds., *The scholar revolutionary: Eoin MacNeill, 1867–1945 and the making of the new Ireland* (New York, 1973).

[15] R. Foster, *Modern Ireland 1600–1972* (London, 1988); J. J. Lee, *Ireland 1912–1985: Politics and society* (Cambridge, 1989); R. English and G. Walker, eds., *Unionism in modern Ireland: new perspectives on politics and culture* (Dublin, 1996).

judicial bench',[16] the action of unionists in Ulster is unsurprising. In September 1912, 250,000 people signed a Solemn League and Covenant which declared:

Being convinced in our consciences that Home Rule would be disastrous to the material well-being of Ulster as well as the whole of Ireland, subversive to our civil and religious freedom, destructive of our citizenship, and perilous to the unity of the Empire [we] do hereby pledge ourselves to stand by one another in defending for ourselves and our children our cherished position of equal citizenship in the United Kingdom and in using all means which may be found necessary to defeat the present conspiracy to set up a Home Rule parliament.[17]

The establishment of an 'unofficial' army, the Ulster Volunteer Force (UVF) in January 1913 enhanced this covenant. The UVF was partly formed to quell violent elements within Ulster unionism but paradoxically to also protect the union through armed resistance if necessary. The force quickly enlisted about 100,000 supporters. Derived from a variety of drilling parties congregating in Orange Halls to oppose Home Rule, the force 'had the appearance of an efficient fighting unit, with a distinguished array of retired army officers. A private army ruled in Ulster with the acquiescence of the state.'[18] In April 1914 about 25,000 firearms and 3 million rounds of ammunition were imported from Germany to Larne, County Antrim to equip this army.[19] This act, Foster claims, was 'both a brilliant publicity coup and an open challenge to the government'.[20] Despite this challenge, John Redmond, leader of the Irish Parliamentary Party at Westminster, maintained his confidence that Home Rule was imminent. He did not, however, enjoy the full backing of nationalist opinion and the Irish Republican Brotherhood (IRB) remained active and alert, albeit with their numbers dwindling to less than 2,000 members.[21] The guiding spirit of the IRB was a Belfast Quaker, Bulmer Hobson, an unshakeable republican and key figure in the articulation of the nationalist cause. By the middle of 1913, Hobson had successfully proposed that members of the brotherhood begin drilling in preparation for the founding of a volunteer force. Underlying this move was the initiation at the beginning of the century of an Irish 'Boy Scout' style movement – the Fianna – where youths were schooled in the Irish language and Irish history as well as in drilling, scouting, military manoeuvres and the use of firearms. This organisation, founded in 1909, would respond to the call for national volunteering if and when the time arose.

Furthermore, another small but significant military movement emerged in Dublin, where class and the national question were being fused. The great Dublin lock-out began with a tramway strike on 26 August 1913. The strike spread to

---

[16] Foster, *Modern Ireland*, 464.   [17] *Ibid.*, 466–7.
[18] D. Fitzpatrick, 'Militarism in Ireland 1900–1922' in T. Bartlett and K. Jeffery, eds., *A military history of Ireland* (Cambridge, 1996), 383–4.
[19] A. T. Q. Stewart, *The Ulster crisis: resistance to Home Rule 1912–14* (London, 1969), 246.
[20] Foster, *Modern Ireland*, 467.
[21] F. X. Martin, *The Irish Volunteers, 1913–15* (Dublin, 1963).

some of the city's other industries including the coal yards, docks and newspapers. Hostility increased in the city when the police force hard-handedly broke up a meeting of unarmed workers on 31 August. The response of the Employers' Federation on 22 September was to lock out all members of the Transport Workers' Union. Around 25,000 men were out of work.[22] This heightening of tension resulted in a call by Jim Larkin, in November 1913, to establish an Irish Citizen Army, under the command of the Ulster Protestant Captain Jack White. Its recruitment was confined principally to Dublin and to members of trades unions. Although this movement sprang largely from labour conflict, particularly in urban areas, the question of conditions of employment, police handling of the dispute and Ireland's position within the Union all became interrelated.

The national issue and the class one were interlinked. The defence of workers' rights could easily translate into the defence of Irish workers' rights within a larger dominion. The Citizens' Army, although numerically small, played a significant role in the Easter rebellion in 1916. James Connolly, its leader, reputedly claimed in prison prior to his execution: 'The socialists will not understand why I am here. They forget that I am an Irishman.'[23]

Against the background of Edward Carson's UVF, industrial unrest in Dublin and Westminster's hesitant position on Home Rule, the seeds were ripe for the foundation of an Irish volunteer organisation. The first initiative towards the establishment of the Volunteers was the publication of an article by the scholar and cultural nationalist, Eoin MacNeill, entitled 'The North Began.'[24] The title of the article is significant as it derived from a line of a poem written by the Young Irelander, Thomas Davis, entitled 'The Song of the Volunteers of 1782'. The article, however obliquely, made connections between eighteenth-century volunteering and early twentieth-century political ideals. MacNeill outlined the threat posed to Home Rule by the UVF. While believing the activities of Carson and his supporters to be a bluff, MacNeill claimed that Ulster's position was 'the most decisive move towards Irish autonomy that has been made since O'Connell invented constitutional agitation'.[25] Treating the UVF as a Home Rule movement, rather than a unionist movement, MacNeill's piece finished on a more conciliatory note suggesting that the UVF and Irish Volunteers might ultimately unite.[26]

After publication of this article a variety of political interests – the IRB, constitutional nationalists and cultural leaders from the Gaelic League – lobbied MacNeill to organise and found a national volunteering movement. Through a series of tactical and complicated political manoeuvres, from a diverse range of quarters,

---

[22] See E. Larkin, *James Larkin: Irish Labour leader, 1876–1947* (London, 1965); J. W. Boyle, *Leaders and workers* (Cork, 1966); M. Connolly, 'James Connolly: socialist and patriot', *Studies*, 41 (1952), 293–308.

[23] Quoted in N. Mansergh, *The Irish Question 1840–1921* (London, 1965), 242.

[24] E. MacNeill, 'The North began', *An Claidheamh Soluis*, 1 November 1913. This was the Gaelic League's newspaper.

[25] Quoted in Martin and Byrne, *The scholar revolutionary*, 131.

[26] MacNeill, 'The North began'.

MacNeill was persuaded to initiate a national movement. Behind-the-scenes meetings were held to prepare for the launch of the movement, with the objective of reflecting as wide a range of political interests as possible. To this end, two prominent supporters of the Irish Parliamentary Party, Larry Kettle (brother of Professor Tom Kettle, MP) and John Gore agreed to join the organising committee. The objective of the movement was to 'secure and maintain the rights and liberties common to the whole population of Ireland'.[27] To that end, the word 'national' was removed from the title of the organisation. Volunteers were to be recruited by locality and not on a class or religious basis. The only cohort of volunteers who would be enlisted as a distinct grouping were university students.

After numerous preparatory meetings by the Steering Committee, a public gathering was arranged for Tuesday 25 November 1913 at the Large Concert Hall of the Rotunda in Dublin. Here enrolment of the Irish Volunteers began. Ireland's third 'unofficial army' came into being. Although the meeting attracted very large crowds, divisions became obvious between supporters of Larkin's labour movement and some of Larry Kettle's supporters who were seen to be unsympathetic to organised labour. The manifesto of the Irish Volunteers, however, stated:

> The object proposed for the Irish Volunteers is to secure and maintain the rights and liberties common to all the people of Ireland. Their duties will be defensive and protective, and they will not contemplate either aggression or domination. Their ranks are open to all able-bodied Irishmen without distinction of creed, politics or social grade. There will also be work for women to do, and there are signs that the women of Ireland, true to their record, are especially enthusiastic for the success of the Irish Volunteers.[28]

The latter section of the declaration presaged the founding of Cumann na mBan, the women's auxiliary corps of the Volunteers. While there were certainly some disturbances at the general meeting, and despite some quiet reservations by John Redmond, the Irish Volunteers were nevertheless founded. Largely a Catholic movement, which recruited one-sixth of all adult Irish males, a geographical analysis of recruitment reveals that 'Participation was most intensive in mid-Ulster, where the promise of conflict with the Ulster Volunteers was most pronounced.'[29] On the eve of the Great War, then, the Irish Volunteers could boast a substantial membership, many of whom would find themselves before long fighting in the trenches of France and Belgium.

The Irish Volunteers, like the UVF, were in need of both finance and weaponry. Money was raised in London and elsewhere and arrangements were made for the import of arms at Howth in Dublin on 26 July 1914. Although the arms shipment arrived, the killing of three people and the injury of another thirty-eight by the army along Bachelor's Walk in central Dublin aroused considerable public discontent. In the meantime, John Redmond sought to win influence within the volunteering

---

[27] Martin and Byrne, *The scholar revolutionary*, 152. [28] *Ibid.*, 171.
[29] Fitzpatrick, 'Militarism in Ireland', 386. Mid-Ulster would generally include the counties of Armagh, east Derry and east Tyrone.

movement. He succeeded in having twenty-five members of the Irish Parliamentary Party added to the central committee of the Irish Volunteers. While the IRB resisted Redmond's attempt to control the movement, they were outvoted on many issues, and, to their dismay, Redmond effectively seized control of the Volunteers.

The summer of 1914 was a most pleasant one in Britain and Ireland, characterised by sunshine and warmth. For a British public, innocence and a belief in progress was disrupted as 'the Great War was perhaps the last to be conceived as taking place within a seamless, purposeful "history" involving a coherent stream of time running from past through present to future'.[30] For the government during that summer, the only anticipated threat to public order was the situation in Ulster as the final stages of the Home Rule Bill had to be passed through parliament. According to what Fussell describes as 'ironic melodrama', at a Cabinet meeting on 24 July 1914 a map of Ireland was laid out for all members to examine, where: 'The fate of nations appeared to hang upon the parish boundaries in the counties of Fermanagh and Tyrone.'[31] That the fate of millions of Europeans would be dictated by the assassination of Archduke Ferdinand and his consort in Sarajevo was unanticipated.

But the onslaught of war did have significant implications for the crisis in Ireland. While the Home Rule Bill would be placed on the statute book, Asquith made two provisos. The bill would not take effect until the war in Europe ended and special amending legislation would yield the opportunity to make provision for Ulster. The bill was given Royal Assent on 1 September and, though it delighted many nationalists, Lyons rightly observes that 'The Irish problem had been refrigerated, not liquidated. Nothing had been solved and all was still to play for.'[32]

In the interim, however, the major event was the Great War, and recruitment of Irish men and women was profoundly influenced by the alliances developing at home. The Home Front and battle front were deeply interconnected in ways different to the rest of the United Kingdom. With volunteer movements partly trained and certainly motivated, the Great War provided an opportunity for each side to display its political and strategic allegiances. The existence of private armies with a membership of over a quarter of a million people posed both a threat and an opportunity for the Crown on the eve of the war.[33] While the months preceding the war seemed to leave Ireland on the brink of civil war, Curtis observes that 'the outbreak of war among the civilised nations of Europe promoted the view that violence was a legitimate, indeed necessary, means of attaining political ends'.[34] Such an ideology would not be lost on the various factions building up in Ireland.

---

[30] Fussell, *The Great War and modern memory*, 21.

[31] Comment made by Cabinet member John Terraine and quoted in Fussell, *The Great War and modern memory*, 25.

[32] F. S. L. Lyons, 'The revolution in train' in W. E. Vaughan, ed., *A new history of Ireland* (Oxford, 1996), Vol. VI. 144.

[33] Fitzpatrick, 'Militarism in Ireland', 383–84.

[34] L. P. Curtis, 'Ireland in 1914' in Vaughan, ed., *A new history of Ireland*, 180.

The Great War, however, quickly exhausted the British Expeditionary Force, compelling Kitchener, secretary of state at the War Office, to recruit a volunteer army. Recruitment drives were therefore initiated throughout British territories and this call to arms yielded different responses in different parts of the empire. In light of this, John Redmond pledged at Westminster on 3 August 1914 that Ireland would support the war effort of the Allied Powers and that volunteers from the north and south of Ireland would defend the island against invasion by the Central Powers, and thus would free up regular troops stationed in Ireland to go on active service to the continent.[35] Redmond sought to reinforce in the minds of the government that 'Home Rule was fully compatible with a loyalty to Crown and Empire.'[36] This strategy of joint action might have a unifying effect across the country and bring unionists closer to nationalist political thinking. Although the Irish Parliamentary Party gave Redmond its support, the IRB and other republican groups regarded the pledge 'as aiding and abetting the enemy of Irish freedom'.[37] Edward Carson pledged the participation of the UVF in the war effort in a display of loyalty to both Crown and Empire. To some degree though, 'the war had greatly weakened the Unionists' bargaining position, for their patriotism prevented them from renewing their threat of civil war in Ulster'.[38] Their public support for the war effort, however, could also generate positive results in the long run. An article published in a local Ulster newspaper in October 1914 perhaps captures this sentiment: 'ULSTER WILL STRIKE FOR ENGLAND – AND ENGLAND WILL NOT FORGET.'[39] A Belfast linen merchant, writing to Carson at the time, also claimed, 'we must stand with them [British soldiers] and for the Empire now'.[40] With Carson's UVF eager to fight overseas, Redmond made a speech at Woodenbridge, Co. Wicklow in September 1914, urging Irish Volunteers not just to defend Ireland at home but to go 'wherever the firing line extends'.[41] This explicit call to join forces on continental Europe proved to be a decisive moment, splitting the volunteers into Redmond's National Volunteers numbering about 170,000 men, and a splinter group opposed to Redmond's strategy forming a new group – the Irish Volunteers – numbering about 11,000.[42] While one group waged war abroad, the other plotted revolution at home, and as one actively promoted the war effort the other obstinately opposed it. Although Eoin MacNeill remained with the small splinter group of Irish Volunteers, Redmond may have considered that he had all

[35] Foster, *Modern Ireland*; F. S. L. Lyons, *Ireland since the Famine* (London, 1971).
[36] Foster, *Modern Ireland*, 472.   [37] Curtis, 'Ireland in 1914', 177.
[38] D. Howie and J. Howie, 'Irish recruiting and the Home Rule crisis of August–September 1914' in M. Dockrill and D. French, eds., *Strategy and intelligence: British policy during the First World War* (London, 1996), 8.
[39] *Newtownards Chronicle*, 31 October 1914.
[40] Quoted in Jeffery, *Ireland and the Great War*, 16.
[41] D. Gwynn, *The life of John Redmond* (London, 1932), 391–92.
[42] L. O'Broin, *Revolutionary underground: the story of the Irish Republican Brotherhood 1858–1924* (Dublin, 1976).

but eliminated his critics.[43] On the eve of the official recruitment drive in Ireland, therefore, the four illegal 'armies' in existence on the island were variously committed to the war on the Western Front, and the tensions within Irish political opinion would all impact on the way the war would be memorialised.

Recruiting an army in Ireland: the early years

The raising of a volunteer army to support the British Expeditionary Force presented an immense challenge for Kitchener and all countries involved in the war. The poster was the principal means of mass communication for recruitment when newspaper circulation was still largely confined to a literate minority.[44] While initially the government hoped that the voluntary principle would encourage men and women to enlist, it soon became clear that greater stimulus was required and that posters and pamphlets would serve this function. The Parliamentary Recruiting Committee (PRC), which was made up of members of all political parties in the House of Commons, commissioned British recruiting posters. Local party organisations were to mediate the message and the committee's work only ceased with the introduction of conscription in 1916.

The existence of pre-war militarism in Ireland might have created a fertile ground for recruiting the hearts and minds of young men. As Fitzpatrick has observed, 'During 1913 and 1914, through an extraordinary outburst of mimetic militarism, a large proportion of Irish adult males began to train, dress and strut about in the manner of soldiers.'[45] Although it is notoriously difficult to estimate the precise numbers of Irish men and women who volunteered, it is now estimated that about 50,000 men automatically transferred from the private armies, especially the UVF and National Volunteers. A further 80,000 were voluntarily recruited and together with the existing Irish servicemen (including regulars, reservists, special reserves and naval ratings and officers), the total number of Irish men in the wartime forces came to about 210,000.[46] The number of women who volunteered for the auxiliary forces remains as yet unknown. The pace of recruitment varied over the course of the war. While the number of volunteers held a steady flow for the first five months of the war, numbers enlisting declined in 1915. After the Easter Rising in 1916, the task of recruiting in Ireland proved immensely difficult and the numbers enlisting declined to a trickle.[47] Together these men, however, formed

---

[43] Lyons, 'The revolution in train'.   [44] Rickards, *Posters of the First World War*.
[45] Fitzpatrick, 'Militarism in Ireland', 383.   [46] *Ibid.*
[47] T. Bowman, 'Composing divisions', *Causeway*, 2 (1995), 1, 24–9. According to Fitzpatrick's (1996) research the numbers enlisting are as follows: 44,000 in 1914; 46,000 in 1915; 19,000 in 1916; 14,000 in 1917 and less than 11,000 in 1918, 'Militarism in Ireland' (388). For analyses of recruitment to particular regiments see T. Denman, 'Sir Lawrence Parsons and the raising of the 16th (Irish) Division, 1914–15', *Irish Sword*, 17, 67 (1987), 90–104; Denman, 'The 10th (Irish) Division 1914–15: A study in military and political interaction', *Irish Sword*, 17, 66 (1987), 16–25; N. Perry, 'Nationality in the Irish infantry regiments in the First World War', *War and Society*, 12 (1994), 65–95; P. Callan 'Recruiting for the British army in Ireland during the First World War', *Irish Sword*, 17, 67 (1987), 42–54.

the backbone of the three New Army divisions – the 10th Irish, 16th Irish and 36th Ulster divisions. The regional name of the latter, secured through persuasion by Carson, signals the delicate balance of loyalty to the war effort in Ireland and the omnipresence of local considerations in the articulation of that commitment.

Motivating men: war propaganda

Each country involved in the war used posters, pamphlets and public addresses to mobilise volunteers and to influence public opinion. Although support for the war in Britain was widespread, posters were one of the primary means of communication and became the locus for the recruiting campaign.[48] As historical documents, posters offer a different insight into how the war was perceived than other types of accounts. Darracott claims that 'Our idea of the First World War is darkly coloured by our knowledge of the tragedy of the battlefields. Posters can give some idea of the flavour of the period as civilians experienced it.'[49] In addition, posters reveal something of the economic and political history of the war as they can indicate different stages in temporal and spatial terms of a country's overall experience of war. They represent official strategies towards recruitment. They are also indicative of attitudes towards munitions work, the food economy and the health needs of combatants.[50] Similarly, posters can reveal some of the contours in the psychology of war and the motivations to serve. While today we are sensitive to the mechanisms underlying advertising campaigns and the semiotic systems that they employ, the reading of the visual syntax of posters and pamphlets may have been quite different in 1914. The poster then becomes an important source for understanding the cultural and political discourse that was regularly deployed to entice young men to enlist.

The PRC specially commissioned posters and their work only ceased with the advent of conscription. While the PRC was prolific in output during the war, the quality of posters from an artistic viewpoint was often suspect. Of a Berlin exhibition of British posters in 1915 a German newspaper reported: 'The exhibition is a great material success, notwithstanding the general disappointment with the poor and inartistic designs.'[51] Whilst ridiculing the enemy's efforts either on the battlefield or in poster design was an essential part of war propaganda, commentators concede that the PRC produced 'a series of posters that was eventually to scrape the bottom of the barrel of persuasion. As graphic art – even British graphic art – it was outstandingly undistinguished. As propaganda it was often painfully inept.'[52] In contrast, French war posters were of much greater artistic merit. At an exhibition of patriotic posters in 1914–20 held at the Sewall Art Gallery, one commentator observed that if the golden age of French poster art was the Belle

[48] Rickards, *Posters of the First World War*.
[49] J. Darracott, *The First World War in posters* (London, 1974), ix.
[50] Hardie and Sabin, *War posters*.
[51] Quoted in Hardie and Sabin, *War posters*, 35.   [52] Rickards, *Posters of the First World War*, 10.

Époque, the silver age was certainly the Great War.[53] As posters they were seen to 'have a vigour and emotional impact that raise them far above their [British] counterparts'.[54] This is partly due to the fact that 'in the figures there is nothing of English photographic precision, nothing of Germany's force and brutality, but always a note of intense sympathy of something subtly human'.[55] While some of these commentaries may reflect the national stereotypes of the day, my concern is not with placing posters within the canon of graphic art but with examining them as part of a discourse on war. In all combatant states posters were employed as part of a larger propaganda system and they regularly masked the reality of trench warfare.[56] As one American poster artist commented, 'The game of war has its horrible side, but it is not advisable to look upon that side in a poster.'[57] While death and mutilation occasionally surfaced in posters, in general they were used to appeal to what were seen as the more honourable motives for enlistment, rather than to exploit the brutality which war necessarily entailed.

Interpreting posters

War recruitment posters form part of the symbolic system of a military conflict and they are produced, circulated, received and negotiated for and by diverse sets of audiences. The academic study of symbols by geographers has received heightened attention over the past twenty years.[58] The initial study of symbolic signs derived much from the work of de Saussure's structural framework.[59] Treating language as a system of signs he embedded his analysis within a semiotic triangle of signifier, signified and referent. While the science of reading signs through linguistic analysis has found some favour and has advanced our understanding of signs as more than mimetic representations of reality or the product of the intentions of the author, critics such as Bakhtin have pointed out that signs are not neutral elements within a single linguistic structure but they are themselves historically and socially constituted. Consequently, the meaning of a symbolic system can only be understood by 'investigating its varied history, as conflicting social groups, classes, individuals and discourses sought to appropriate it and imbue it with their own meanings'.[60] Post-structuralists, therefore, have moved beyond the constraints

[53] J. L. Jackson, *French patriotic posters 1914–1920* (Houston, TX, 1989).
[54] Rickards, *Posters of the First World War*, 25.
[55] Hardie and Sabin, *War posters*, 18.   [56] Hillier, *Posters*.
[57] Quoted in Rickards, *Posters of the First World War*, 35.
[58] See for instance, D. Meinig, ed. *The interpretation of ordinary landscapes* (Oxford, 1979); D. Cosgrove, *Social formation and symbolic landscape* (London, 1984); Cosgrove and Daniels, eds., *The iconography of landscape*; Duncan, *The city as text*; Harley, 'Deconstructing the map' in Barnes and Duncan, eds., *Writing worlds*, 231–47; Duncan and Ley, eds., *Place/culture/representation*; J. Ryan, *Picturing empire* (London, 1997); R. Schein 'The place of landscape', *Annals of the Association of American Geographers*, 87 (1997), 660–80.
[59] F. de Saussure, *Course in general linguistics* (London, 1960); R. Barthes, *The elements of semiology* (London, 1967); *Mythologies* (London, 1972); *The pleasure of the text* (New York, 1975).
[60] Quoted in T. Eagleton, *Literary theory: an introduction* (Minneapolis, 1983), 117.

of early literary structuralism by claiming that meaning itself is not necessarily directly present in a sign but is mediated through layers of signification that are geographically and temporally contingent. There is a geography of hermeneutics just as much as there is a history. By challenging the binary oppositions of structuralism, post-structuralism attempts to 'demonstrate how one term of an antithesis secretly inheres within the other'.[61] Although emerging from literary theory and the analysis of literary texts, the analysis of symbols has expanded to consider other spheres of interpretive representation such as painting, advertisements and landscapes.

Although the early work of Roland Barthes shared the structuralist emphasis of de Saussure, his later work moved towards an approach to representation which was looser than the strict constraints of semiology and transcended the limits to interpretation permitted by the semiotic triangle. Barthes encapsulated this expanded sphere of interpretation through the concept of myth. Underlying the language of signs, he claimed, lay a metalanguage of myth which conveys a deeper and global message. Barthes used the images and texts of advertising to illustrate his approach, yet some of the shortcomings of his method have been observed by geographers. An overemphasis on linguistic signs, the treatment of semiotics as a static, closed system of representation, the ahistorical focus of the method and the attempt to bring interpretation totally into the realm of scientific inquiry have all led geographers dealing with the meaning of texts to 'include other cultural productions such as paintings, maps and landscapes, as well as social, economic and political institutions'.[62] Recruitment posters can be included as symbolic texts which exist within a larger discourse about the nature of modern warfare and the duty of individuals to serve their country. Posters, then, can be seen as an intrinsic part of the propaganda and advertising for the recruitment of individual soldiers but they also form part of the larger narrative on the cultural meaning of war. They act as mediating texts between the individual and the larger body politic. As such, their reading becomes more than just a textual analysis of a single message but rather a window into a wider ideological and material world. As Foucault reminded us, 'semiology is a way of avoiding its violent, bloody and lethal character by reducing it to a calm Platonic form of language and dialogue'.[63]

Taking the analysis of symbols towards the mythological and treating the texts of advertising as part of a wider discourse of war overcomes some of these constraints of rigid semiotics. It also heightens the possibility of an alternative reading of war. In posters the language of the written text and the visual image combine to naturalise a set of relationships between war as an act of killing and war as a moral discourse which legitimates killing. The ways in which these messages are translated, however, vary from poster to poster, and the reception of the messages by the audience is always mixed. The spaces of production and consumption

---

[61] *Ibid.*, 133.
[62] T. Barnes and J. Duncan, 'Introduction' in Barnes and Duncan, eds., *Writing worlds*, 5.
[63] M. Foucault, *Power/knowledge* (Brighton, 1980), 115.

intersect. In the following analysis, although the dominant tropes of duty visible in war propaganda will be highlighted, the reading of these texts against complex attitudes towards Irish participation in the war will be considered. If this were not the case, all eligible men of good health and appropriate age would have enlisted. In reality they did not.

Irish war propaganda

Initially posters in Ireland had no distinctly Irish content. The images employed around Britain were also used to relay the message in Ireland. With the aim of using everything at their disposal to induce enlistment, propagandists 'ran the gamut of all emotions which make men risk their lives'.[64] Early posters tended to focus on loyalty to the empire and the crown:

Lord Kitchener says 'The time has come, and I now call for 300,000 recruits to form new armies'. God save the King

The message was clear and straightforward, denoting the need for 300,000 new troops and this is an inevitable request in wartime circumstances where king and country must be defended. While this message may have had some appeal in Ireland, especially among unionists, the war needed to be domesticated more explicitly to the local political context to find wider attraction.

It was agreed in early 1915 to establish a separate recruiting board, the Central Council for the Organisation of Recruiting in Ireland (CCORI). Its central office was on Great Brunswick Street in Dublin. The honorary president was Ireland's lord lieutenant, while the council itself was under the chairmanship of the city's lord mayor. John Redmond and the Irish Parliamentary Party (IPP), however, were at the forefront of the recruitment campaign, pledging to create 'an atmosphere favourable to recruiting' among nationalists.[65] Large sections of the Irish press and key leaders of the IPP (including John Dillon and Joseph Devlin) offered their support to CCORI. The council's remit was as follows: to form and assist local recruiting committees in every urban and rural district in Ireland; to assist county recruiting committees to coordinate and develop recruiting drives; to liaise between the central office and the local committees. A travelling recruiting officer would also be established, often accompanied by speakers from central office. The travelling recruiting officer would focus on places where there were no locally organised recruiting committees or places remote from railway stations.[66] This organisation, as well as the two which succeeded it, the Department for Recruiting in Ireland (October 1915) and the Irish Recruiting Council (May 1918), produced

---

[64] C. Haste, *Keep the home fires burning: propaganda in the First World War* (London, 1977), 52.

[65] T. Denman, *Ireland's unknown soldiers: the 16th Irish Division* (Dublin, 1992), 30.

[66] *Pamphlet*, Central Council for Recruiting in Ireland (Trinity College Dublin: Recruiting leaflets relating to European War, 1914–18, OLS L-1-540 Nos. 1–16), n.d.

a series of large-format posters and pamphlets which circulated throughout Ireland in the war years, and it is to them that I now wish to turn.[67]

Recruitment pamphlets

The recruitment pamphlets that appeared in support of the war frequently employed the voices of local Irish political leaders, officers at the Front, church leaders or army chaplains. In an appeal made by John Redmond MP in February 1916 the legitimacy of the war was strongly promoted: 'a just war, provoked by the intolerable military despotism of Germany; that it was a war in defence of the rights and liberties of small nationalities' (Figure 1). Although working under the flag of empire, Redmond emphasised that there was a 'distinctively Irish army composed of Irishmen, led by Irishmen'. He claimed that the sacrifices made north and south of the island 'will form the surest bond of a united Irish Nation in the future'. The war was being domesticated for an Irish nationalist constituency, which combined both a moral duty to defend against a despot and equally appealed to the idea that political unity on the island could be secured through such defence. Constitutionalists consistently mirrored Britain's broader claim of the righteousness of the war in the defence of small nations. The moral principle of nationality was being invoked but one which emphasised the multinational character of Britain's empire. Bonds of reconciliation, Christian brotherhood and mutual recognition cemented that empire and for Redmond the Home Rule Bill underlined that characterisation of empire.[68] Moreover, an emphasis on Christian brotherhood endeared the Catholic Church to the war effort. Religion and Godliness combined stimulated the Catholic hierarchy's attitude towards German attacks on churches and cathedrals in Belgium. One priest claimed that 'Militarism is the gospel of force. It is a negation, therefore, of Christianity.'[69] Denominational differences could be set aside for higher moral codes while the continuous distinctiveness of Irish religiosity could also be stressed. One propagandist claimed: 'The Irish are the most

---

[67] The most comprehensive collection of Irish recruitment posters is held in three volumes in the Department of Early Books, Trinity College Dublin (Call nos. 22.7.27-29, Papyrus Case 16). While it is not known just how many different posters were produced and thus whether this is a complete set, it does contain a wide-ranging sample. The set comprises 203 posters, two-thirds of which were large enough for public display and the smaller ones may have been used inside recruiting stations. The print runs for these posters range from 250 to 40,000 and it is estimated that 2 million were produced in Ireland in total. Most posters are published in colour with green or red being the predominant colours. About 75 per cent of the posters reflected some Irish content, especially in 1915 when the recruitment drive was at its most intense. For a fuller discussion see M. Tierney, P. Bowen and D. Fitzpatrick, 'Recruiting posters' in D. Fitzpatrick, ed., *Ireland and the First World War* (Dublin, 1986), 47–58.

[68] J. Ellis, ' "The methods of barbarism" and the "rights of small nations": war propaganda and British pluralism', *Albion*, 37 (1998), 49–75.

[69] Quoted by J. Ellis, 'The degenerate and the martyr: nationalist propaganda and the contestation of Irishness. 1914–1918', *Eire-Ireland*, 35 (2000), 7.

APPEAL FROM

JOHN REDMOND, M.P.

To the People of Ireland.

AT the very commencement of the War I made an appeal to the Irish people, and especially to the young men of Ireland, to mark the profound change which has been brought about in the relations of Ireland to the Empire, by whole-heartedly supporting the Allies in the field.

I pointed out that, at long last, after centuries of misunderstanding, the democracy of Great Britain had finally and irrevocably decided to trust Ireland, and I called upon Ireland to prove that the concession of liberty would, as we had promised in your name, have the same effect in our country as in every other portion of the Empire, and that Ireland would henceforth be a strength, instead of a weakness.

I further pointed out that this was a just war, provoked by the intolerable military despotism of Germany ; that it was a war in defence of the rights and liberties of small nationalities; and that Ireland would be false to her history and to every consideration of honour, good faith, and self-interest if she did not respond to my appeal.

I called for a distinctively Irish Army composed of Irishmen, led by Irishmen, and trained for the field at home in Ireland.

I acknowledge with profound gratitude, the magnificent response the country has made.

For the first time in history, we have to-day a huge Irish army in the field. Its achievements have covered Ireland with glory before the world, and have thrilled our hearts with pride.

North and South have vied with each other in springing to arms, and please God, the sacrifices they have made side by side on the field of battle will form the surest bond of <u>a united Irish Nation</u> in the future.

Figure 1 John Redmond pamphlet

religious soldiers in the British Army; and it is because they are religious that they rank so high among the most brave.'[70] This pamphlet then, whose fundamental aim was to encourage men to fill reserve battalions in defence of Ireland, can be seen within Britain's broader narrative of the moral duty to serve. The message to join the reserves, simple at one level, was therefore a complex text in its articulation of the relationship between Ireland and Britain, Home Front–battlefront, north–south, small nations and Ireland.

Other pamphlets were more generalist in content, making no particular reference to Ireland but relying on conventional gender stereotypes such as the following: 'Have you got a mother, a sister, a girl or a friend worth fighting for?' (Figure 2). Likening the fate of British women to those in Belgium and France, the pamphlet concludes with the ultimate guarantee of masculinity – 'Thank God I too was a man.' While commentators have noted that early poster designs were 'cheap in sentiment' and seized upon 'childish and vulgar appeals to patriotism',[71] pamphlets also deployed the same textual messages. But, as Fussell reminds us, 'war took place in what was, compared to ours, a static world, where the values appeared stable and where the meanings of abstractions seemed permanent and reliable. Everyone knew what Glory was, and what Honour meant.'[72]

While Fussell alerts us to the dangers of presentism in historical analysis, the stability of meaning he suggests existed in this period did not go at the time totally unchallenged. The suffragist movement in the period before the war was taking gender inequality seriously and disrupting accepted values.[73] Labour unrest across Europe in the year prior to the war also underlined the fact that conventional class relations were also experiencing challenges. In an Irish context in particular the meaning of terms like honour and loyalty were not so stable. Seemingly universal principles like duty had to be localised, mediated and translated to specific contexts, and this was achieved by conveying these meanings through local discourses rather than through universalising ones. The signifying system was operating therefore within a broader series of tropes and the tension between the sign and its local translation was the tension between, on the one hand, loyalty to an empire by unionists, and efforts by nationalists to achieve Home Rule or independence on the other. The syntax of recruitment literature recognised and accommodated these tensions and the instability of meaning attending to a sign, noted by Barthes, is evident in Irish recruitment propaganda. As Eagleton observes: 'The "writable" text, usually

---

[70] M. Mac Donagh, *The Irish at the front* (London, 1916), 104.

[71] Hardie and Sabin, *War posters*, 2.

[72] Fussell, *The Great War and modern memory*, 21.

[73] C. Rover, *Women's suffrage and party politics in Britain, 1866–1814* (London, 1967); R. Strachey, *The cause: a short history of the women's movement in Great Britain* (Bath, 1974); M. Pugh, *Women's suffrage in Britain, 1867–1928* (London, 1980); C. Law, *Suffrage and power: the women's movement, 1918–28* (London, 1997); N. Dombrowski, *Women and war in the twentieth century: enlisted with or without consent* (London, 1999).

" There's a neater, sweeter maiden,
In a cleaner, greener land."

Have YOU got a mother, a sister, a girl or a friend worth fighting for?

Do you realise that they may share the fate of the daughters of France and Belgium?

And having realised it, don't you think it worth your while to go and don Khaki so that you can prevent such a fate as that?

Never mind whether they ship you "Somewhere East of Suez" or "Somewhere in France" or whether they put you on Home Defence.

Do your bit and when it's all over and "Tommy" comes home you'll be able to say

"THANK GOD I TOO WAS A MAN."

Wt. 1788. 50,000. 6/16. P.P.D.

Figure 2 Recruitment pamphlet

a modernist one, has no determinate meaning, no settled signifieds, but is plural and diffuse', although he acknowledges too that 'If there is any place where this seething multiplicity of the text is momentarily focused, it is not the author but the *reader*.'[74] And the readers in Ireland, although diverse, did fall into a finite number of political camps.

Political and religious leaders added the weight of their office to the recruitment campaign. Reverend John McMullan CP, a provincial of the Passionate Order in Mount Argus in Dublin, made the following plea after his return from the trenches:

This is not only a crisis in the history of the world, but a crisis in the history of Ireland, for if the Allies lost, Ireland would be under the heel of Germany. Some people said the

[74] Both quoted from Eagleton, *Literary theory*, 138.

Germans had promised to found an Irish Republic; but what were German promises worth to Belgium when they violated her neutrality.[75]

With respect to the impact the war would have on local politics, he claimed that:

They might have had misunderstandings in the past on religion in Ireland, but the men who were fighting and dying side by side for the same cause in France would come back as brothers, and one of the grandest results of the war would be the unity of Irishmen for nationality, freedom and brotherhood.[76]

The unity presupposed by this statement masked the real divisions that continued to characterise Irish politics at the time, but it did provide a legitimating vocabulary to the rewards of war. Soldiers were also lobbied by the Central Recruiting Council to write open letters to the public encouraging enlistment. Stephen Gwynn, nationalist MP for Galway city, who had enlisted as a private in an Irish regiment but rose to the rank of captain in the Irish Brigade (16th division), published a pamphlet in July 1915 in which Irish men were encouraged to listen to their conscience:

If any ask, why it is Ireland's war, there is a plain answer. It is a war in defence of justice and liberty. The case of Belgium alone suffices. One of the most monstrous wrongs in history lies there to be redressed. We, as Irishmen, cannot stand idly by and see a nation brutally trampled into servitude, and its heroic resistance made of no account.[77]

Gwynn strengthened his case by emphasising the practical consequences of war:

But it is not only the honour which compels us. Germany holds Belgium, and so long as she holds it the empire of which Great Britain is the centre can never be at rest. Unless the Germans are driven out of Belgium we may look forward to an endless sacrifice of our personal freedom. We shall be driven into conscription.[78]

Thus war could impact on personal liberties as well as larger political ideals. With respect to the question of social class (which animated the thinking of labour and trade unionists in Ireland) Gwynn emphasised the levelling effect of war on class divisions:

No man is too good to carry a rifle in it [the army], and for an educated citizen the most honourable position is, perhaps, a place in the ranks. The best bred can find companies in several Irish regiments where his comrades will be as well bred as he.[79]

[75] Pamphlet, by Reverend John McMullan CP, *Irish soldiers at the front* (Trinity College Dublin: Recruiting leaflets relating to European War, 1914–18, OLS L-1-540 Nos. 1–16), n.d., p. 8.
[76] *Ibid.*, 8.
[77] Pamphlet, *An open letter from Capt. Stephen Gwynn MP to the young men of Ireland*, n.d. (Trinity College Dublin: Recruiting leaflets relating to European War, 1914–18, OLS L-1-540 Nos. 1–16).
[78] *An open letter from Capt. Stephen Gwynn MP to the young men of Ireland.*
[79] *Ibid.*

Pamphlets thus sought to underline the righteousness of the cause, the duty of Irish men to serve, the political benefits that might accrue to Ireland in exchange for loyalty and the unifying effect of war on political and class divisions. Although similar in tone to pamphlets used elsewhere in the United Kingdom, the use of local voices to mediate the message, and the incorporation of local circumstances into these texts helped to establish Ireland's particular role in defending the Allied case. Appeals to the destruction of small nations, the possibility of conscription and the moral imperative to serve, delivered by the representatives of moral authority, Catholic clergymen, gave added weight to the state's appeal for men to enlist. These messages, however, were not uniformly received.

Running parallel to the overall recruitment campaign was the campaign of sepa-ratists who portrayed enlistment as an unpatriotic act. This anti-recruitment propa-ganda found expression in the newspapers of radical organisations such as the IRB (*Irish Freedom*), Sinn Féin (*Éire* and *Sinn Féin*), the Irish Volunteers (*The Irish Volunteer*), and the Irish Transport and General Workers' Union (*Irish Worker*). To-gether these papers represented an altogether different perspective on the war and a different interpretation of Irish nationality. *Irish Freedom*, in particular, claimed the war to be England's, motivated 'not for the cause of religion or civilisation but for the cause of England's great God-Markets'.[80] Rather than protecting the rights of small nations, the same paper claimed, England had deployed the vocabulary of righteousness 'to keep small Nationalities in subjection'.[81] The Irish recruit, therefore, was a representation of degeneracy, trading his moral authority for the king's shilling. Constitutional politicians, promoting this course of action, were similarly corrupt in their exchange of moral principle for 'the power of English gold'.[82] This was confirmed, according to separatists, through the £400 'bribe' Irish MPs accepted to recruit.

Anti-recruitment campaigners similarly deployed the language of religion to present their case. They labelled John Redmond a 'Judas' and his followers as 'po-litical Esaus'.[83] Rejecting the constitutionalists' definition of nationality, which accommodated a sense of a collective British cause with a distinctive Irish compo-nent where each 'nation-in-arms' could stand beside one another under a common flag of empire, separatists claimed such ideological manoeuvring represented 'na-tional apostasy'.[84] Irish nationality was being degenerated by being diluted by an English one. Supporters of recruitment were at best 'West Britons', but unlike the English man, they suffered from a schizophrenia. They were being neither truly Irish nor English, neither truly loyal nor disloyal. For separatists, any claim to a kind of dual nationality or identity was illogical and incoherent. The attempt to promote this illogicality through the lexical gymnastics of recruitment campaign-ers was to condemn the Irish soldier to an untenable position. This position was

[80] *Irish Freedom*, December 1914.   [81] *Irish Freedom*, November 1914.
[82] *Irish Freedom*, October 1914.   [83] *Irish Freedom*, September 1914.
[84] Ellis, 'The degenerate and the martyr', 19.

represented graphically through the body of the dead soldier – a rotten, unburied and unidentifiable corpse lying on the field of battle devoid of identity or meaning. Indeed the dead soldier became the embodiment of the vacuousness of the recruiter's mission in anti-recruitment texts.

For the separatist wing of Irish political opinion, then, the constitutionalist and his army of recruits was a symbol of a degeneracy in Irish political culture and this degeneracy was confirmed by the lack of political support for separatists in the build-up to the Easter Rising. Ellis tellingly suggests that 'The separatists did not want to fight for the national vision expressed by the majority but to transform that vision.'[85] Part of their mission to transform it was to oppose the war openly in their writings. However, after the Easter Rising, a more conciliatory tone entered their vocabulary, as Sinn Féin's expanding support base included the families of many serving soldiers.

Ironically, in common with the imagery of sacrifice and resurrection relayed through the empire's propaganda machine, for separatists too the Easter Rising revivified the idea of national redemption through martyrdom and blood sacrifice. Death, in the proper circumstances, could act as a spiritual triumph as well as a political one. The writings of anti-recruitment campaigners no doubt had some influence on men's attitudes towards enlistment, particularly after the Rising. In the early years of the war, however, constitutionalists and official recruitment agencies dismissed anti-recruitment opinions as the work of a handful of cornerboys without authority or support. For unionists, such expressions of disloyalty undergirded their suspicion of constitutional nationalist support for the empire in general and the underlying, persistent existence of republican dissent in particular.

Drawing support: recruitment posters

Visual representations of the war were employed by each combatant state in its recruitment literature. While the style of representation varied in terms of artistic codes, the messages that the images conveyed shared the common objective of enhancing enlistment. They were one of a suite of texts that mediated a sense of the war to those who remained at home. As Jay has commented with respect to visual images, 'what is "seen" is not a given objective reality but an epistemological field constructed as much linguistically as visually'.[86] Posters did not, then, represent the objective reality of the war being waged in Europe; they provided a lens for popular interpretations of the conflict for particular purposes.

A number of dominant themes emerge from the study of Irish war posters. Before 1915, posters did not have a distinctly local flavour and even after the establishment of Irish recruitment councils, some posters employed those generic motifs that characterised posters throughout Britain. Such posters usually framed

[85] *Ibid.*, 20.
[86] M. Jay. 'In the empire of the gaze: Foucault and the denigration of vision in twentieth century French thought' in D. C. Hoy, ed., *Foucault: a critical reader* (Oxford, 1986), 182.

Figure 3 Recruits required

a soldier in uniform, regiment details, age requirements, pay and conditions and the name of the local recruiting station. Figure 3 represents this genre: a sturdy, upright soldier is shown in marching stance. Similarly, specific units of the armed forces – the Tank Corps (Figure 4), the Royal Navy (Figure 5), the Royal Air Force (Figure 6) – employed the image of a man in peak physical condition wearing a clean, well-pressed uniform and prepared for, but never actually engaged in, battle. While the recruitment age was between 18–41 years, the pictorial representation of soldiers rarely employed images of young men. The fact that youthful soldiers were being killed in vast numbers cannot be gleaned from looking at these posters. Although the language is immediate, direct and without significant

Figure 4 The Tanks Corps

embellishment, the visual image suggests that the work is clean, safe and produces sturdy, well-groomed men. In Barthes' terms 'the excellence of the product is announced'.[87] But as he notes in relation to commercial advertising, denotation helps to develop certain arguments, in short to persuade; but it is more likely . . .

that the first message serves more subtly to *naturalise* the second: it takes away its interested finality . . . it substitutes the spectacle of a world where it is *natural* to buy . . . the commercial motivation is thus found not so much masked as *doubled* by a much broader representation, since it puts the reader in communication with the great human themes.[88]

[87] R. Barthes, 'The advertising message' in R. Barthes, *The semiotic challenge* (Oxford, 1988 [1963]), 174.
[88] *Ibid.*, 176.

Figure 5 The Navy wants men

The interested finality in this case, that men in the armed forces are trained to kill or be killed is obscured by the representation of soldiering as a healthy, well-paid occupation.

War posters frequently appealed to what were seen as universal principles of duty and masculinity placed within specific national and imperial contexts. Four ways of representing these themes are evident in Irish recruitment posters. First, there was an appeal to Irishmen living in the countryside to protect their agricultural homeland from foreign invasion (Figure 7). In this instance the homeland is encapsulated as an idyllic rural landscape. The message, 'Farmers of Ireland Join Up and Defend your Possessions', is framed with a unit of marching troops

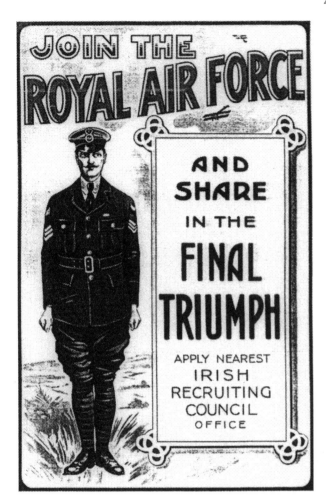

Figure 6 Join the Royal Air Force

making its way through the countryside and being waved at by the women and children of the farmsteads. The juxtaposition of a large two-storey farmhouse and a single-storey thatched cottage, while highlighting class differences in the rural economy, also serves to convey the common cause that war represents. In this context, differences in rural wealth and prosperity are superficial dimensions of identity. The pastoral landscape, which both big farmhouse and thatched cottage helped to create, is of more significance than the individual representations of wealth. The land must be defended irrespective of one's class position. The pictorial representation of a rural landscape naturalises the message to enlist and to

Figure 7 Farmers of Ireland

defend national territory by representing it in pastoral terms. The fact that agrarian unrest and the land question underpinned so much of late nineteenth-century nationalist politics[89] is subtly disguised in this poster.

Although recruitment in urban areas consistently outnumbered that of rural areas in Britain and Ireland, recruiters did see the value of representing the pastoral idyll as a trope of a land worth defending. For both the Irish and the British sense of national identity the rural landscape was a dominant leitmotif. Rural themes are also displayed in Figures 8 and 9, both of which have the same picture although with different captions. In these two examples, the rural, the historical and the religious are interspersed in a single image. In contrast with the previous examples the rural landscape is foregrounded by a robust and humble farmer ploughing his plot of land. In the background is the figure of Saint Patrick, patron saint of Ireland, suspended in mid-air, bearded, carrying a crosier and standing beside a cathedral. The farmer, seeing the apparition, removes his cap in modesty to the higher authority. The caption reads 'Can you any longer resist the call?' With the moral authority of religious insistence, the farmer is persuaded to answer the call. The war, thus, did not derive its legitimacy from human authority alone: it was underscored by the sanction of the blessed authority of the figure attributed with bringing Christianity to Ireland.

[89] Garvin, *The evolution of Irish nationalist politics.*

Figure 8 Can you any longer resist the call?

Theological justification for the war underpinned many states' legitimating theses. In the context of Canada, Vance has pointed out that 'Just as Jesus had given his life so humanity could survive, so too did the soldiers offer their lives for humanity. In this theology, each death was an atonement, each wound a demonstration of God's love, and each soldier a fellow sufferer with Christ.'[90] The universalising pronouncements of Christianity were localised for particular cultural communities and the discourse of war mediated differentially. Irish religious leaders made pronouncements: the Catholic primate, Cardinal Logue, condemned 'the barbarism of the Germans in burning Rheims Cathedral', while the *Cork Free Press* intoned that

[90] Vance, *Death so noble*, 36.

Figure 9 The Isle of Saints and Soldiers

' "Louvain and Rheims" alone are cries which would stir the blood of Catholic Irishmen.'[91] In Figure 9 an identical picture is accompanied with the caption, 'The Isle of Saints and Soldiers', which is a reworking of the cliché 'the island of Saints and Scholars'. The juxtaposition of militarism and religious iconography, of course, was not confined to the war of 1914. Ironically, this marrying of war and religion had found expression in earlier military conflicts, as well as among the various pre-war unofficial armies in Ireland. Saintliness and soldiering could cohabit especially in a world where martyrdom for one's religious faith had strong cultural resonances.

While ethnic stereotyping of the enemy was employed regularly as a common motif in recruitment posters, stereotyping of the 'native' also proved useful as a

---

[91] Both quoted by Jeffery, *Ireland and the Great War*, 12.

means of exerting pressure to enlist. The representation of 'other' has received widespread scholarly attention since the publication of Said's treatise on Orientalism,[92] and the image of the Irish in Victorian Britain has been modified since Curtis' early work, *Apes and angels: the Irishman in Victorian caricature*, published in 1971.[93] This analysis focused on the images of Irish people as bestial sub-humans, images which were based on crude pseudo-Darwinian interpretations of race. As Jackson has noted, however, 'racism is not a uniform or invariable condition of human nature but, like other ideologies, is firmly rooted in the changing material conditions of society'.[94] In an analysis of cartoon representations of Irish people in *Punch* in the nineteenth century, Foster warns of reducing the depictions to simple racial prejudice.[95] The relationship and representations of Irish people in Britain were ambiguous and did not derive solely from the biological metaphors suggested by Curtis. Religion and social class also played a role and, although negative stereotypes were embedded in a colonial relationship, it was more complicated than a simple 'master–slave' narrative would suggest. Britain's relationship with Ireland was never simply one of a colonial overlord patronising the native in the will to power. It varied across historical period, class and region.

Negative stereotypes thus emphasising the idleness, stupidity and bestiality of the Irish would not enthuse volunteers to fight for the crown.[96] Positive stereotypes therefore would have to be employed, as illustrated in Figure 10. This poster displays an 'Irish Hero', a recipient of the VC, and it uses this image as a synecdoche of Irish 'character'. The Irish soldier – a man from the ranks – defeats ten Germans and appeals to his fellow countrymen to emulate his bravery.[97] Similarly, in Figure 11 a potentially negative image becomes a positive one, as a tired and wounded German soldier holds a banner stating that Ireland's 'old fighting spirit' persists with 'thousands joining the colours'. An appeal to the natural predisposition of Irish men to engage in aggressive behaviour could be transformed into a positive image when that aggression is channelled in the cause of the just war. For Irish separatists, though, the just war was not found in continental Europe, rather it resided at home in the 'National Tradition of Tone and Emmet, depending on the

[92] The most influential writing on this topic is E. Said, *Orientalism* (London, 1979).

[93] L. P. Curtis, *Apes and angels: the Irishman in Victorian caricature* (Newton Abbot, 1971).

[94] P. Jackson, *Maps of meaning* (London, 1989), 132.

[95] R. Foster, *Paddy and Mr Punch: connections in Irish and English history* (London, 1993).

[96] For a discussion of images of Irishmen's behaviour in the army see T. Denman, 'The Catholic Irish soldier in the First World War: the "racial" environment', *Irish Historical Studies*, 108 (1991), 352–65.

[97] Michael O'Leary, a lance-corporal, serving with the Irish Guards during the winter of 1914–15 in France, came under attack as German soldiers moved forward on 25 January 1915. In an attempt to regain position Michael O'Leary captured two barricades in the La Bassée sector of the trenches, captured two enemy soldiers and killed eight more. He was subsequently awarded the Victoria Cross. T. Johnstone, *Orange, green and khaki: the story of Irish regiments in the Great War, 1914–18* (Dublin, 1992).

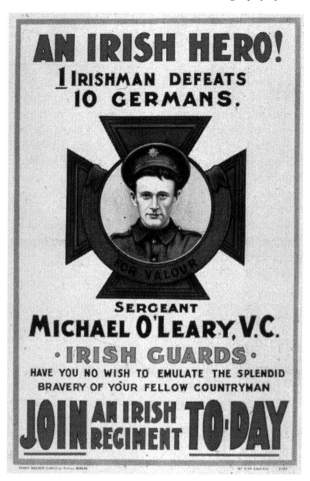

Figure 10 An Irish Hero

inherent righteousness and sanctity of a cause hallowed by the blood of martyrs'.[98]
An appeal to an image of national character is again exploited in Figure 12 where
a soldier and a civilian meet in a landscape replete with symbols of 'Irishness'.
The round tower and ruined church form the background for the image and may
be loosely based on an impression of the early Christian monastic site in Glen-
dalough, Co. Wicklow. The civilian, dressed in dandy attire, perhaps representing
the landed gentry but legitimated through the lens of an 'older' Ireland reflected
in the background, declares 'I'll go too.' The poster's message, that this declaration represents 'The real Irish spirit', one based on loyalty and duty rather than

[98] *Irish Freedom*, November 1914.

Figure 11 Ireland's old fighting spirit

a 'natural' aggressiveness, again contravenes the negative stereotype sometimes attached to the idea of an 'Irish spirit'.

In case the public doubted the availability of additional volunteers a cartographic image of the Home Front and battlefronts in Figure 13 reminds the population of the numbers still available to enlist. Although undated, this war map represents the sites of battle in which Irish regiments played a prominent role, including Gallipoli, Bethune and Givenchy.[99] The situating of Ireland's war effort at specific places and battles would take on even greater significance in the constitution of Ulster unionist's identity surrounding the battle of the Somme. The pictorial connection

[99] Although the map is undated one can speculate, from the places listed on the map, that the poster was produced before the Somme offensive in July 1916.

Figure 12 I'll go too

between the war front and the Home Front exemplified through this mapping metaphor constructs the war not as a collective effort of Allied forces, but as an effort disaggregated along ethnic lines and regional battles to enthuse men to 'answer the call'. The war is mediated locally then because like other texts 'cartography also belongs to the terrain of the social world in which it is produced. Maps are ineluctably a cultural system.'[100] The use of maps in recruitment posters can be seen as part of a larger process of persuasion where, 'The propagandist's primary concern is never the truth of an idea but its successful communication to a public.'[101] Mapping, as well as war art and journalism, was part of this process.

[100] Harley, 'Deconstructing the map', 233.
[101] Quote from Speier in J. Pickles, 'Texts, hermeneutics and propaganda maps' in Barnes and Duncan, eds., *Writing worlds,* 197.

Figure 13 Ireland's War Map

After the war, Lord Northcliffe observed that 'The bombardment of the German mind was almost as important as the bombardment by the cannon.'[102] We could add that bombardment of the mind on the home territory also played a crucial role in achieving victory. The visual representation of Ireland's war fought in eastern France and the Middle East brought these relatively remote places home. Absent from this map are soldiers or German territory but the mapping metaphor transforms the metaphorical into a seeming objective representation of statistical fact. There are 100,000 men available as 'Ireland at the Front Looks to Ireland at

[102] Quote from Lord Northcliffe, *ibid.*, 202.

Home to Answer the Call.' As Barthes has commented, the text accompanying an image 'loads the image, burdening it with a culture, a moral, an imagination'.[103] The imagination stimulated in this war poster is of Irish men serving in far-off lands for the defence of civilisation.

Although in some respects the extraordinary conditions generated by the war challenged orthodox gender conventions, especially with female participation in the workforce[104] and the extension of female social and sexual freedoms,[105] some feminist historians have suggested that the war acted as a 'double helix' in the delineation of gender roles. The metaphor of two entwined strands suggests that 'continuity [lay] behind the wartime material changes in women's lives'.[106] In terms of war posters the use of female imagery was both conventional and radical. The discourse of vulnerability of women in a war context was emphasised while the war itself provided some of the conditions for challenging this image. In Figure 14 a picture of a young woman with a headscarf and angelic face frames the text. The reader is asked, 'Have you any women folk worth defending?' The question mark queries the notion that all women are worthy of protection. The saintly innocence of womanhood represented in the poster, however, confirms that they are. The second caption moves to a more specific national representation of woman: 'Remember the women of Belgium – Join today.' The text serves to re-inforce the stories circulating through Allied propaganda about the treatment of women, particularly Belgian nuns, by German soldiers. This broader discourse centred on the brutality of the enemy and symbolised this belief through the image of the helpless woman. Propagandists' desire to recruit by shame seized upon this image.

In Figure 15 the iconography draws from classical female mythological figures and images of Irish womanhood. The bugling soldier is framed within a harp and beside the harp stands a red-haired, crowned woman, in classical dress and appearing to listen to the bugler's tune. Gendered depictions of graces and virtues have a long history in European art, and female allegory has often been associated with 'the common relation of abstract nouns of virtue to feminine gender in Indo-European languages'.[107] In this poster the classically dressed figure of Hibernia pleads that it is now time to answer the call of Mother Ireland. Again a landscape

---

[103] R. Barthes, *Image, music, text* (New York: Hill and Wang, 1987), 26.

[104] A. Bravo, 'Italian peasant women and the First World War' in C. Emsley, A. Marwick and W. Simpson, eds., *War, peace and social change in twentieth century Europe* (Milton Keynes, 1989), 102–15; Ouditt, *Fighting forces*; C. Braybon, *Women workers in the First World War: the British experience* (London, 1981); J. Elshtain, *Women and war* (Brighton, 1987).

[105] C. Tylee, *The Great War and women's consciousness: images of militarism and womanhood* (London, 1990); D. Riley, *War in the nursery* (London, 1983); D. Campbell, *Women at war with America: private lives in a patriotic era* (Cambridge, MA, 1984).

[106] M. Higonnet and P. L.-R. Higonnet, 'The double helix' in Higonnet *et al.*, *Behind the lines*, 39.

[107] M. Warner, *Monuments and maidens: allegory of the female form* (London, 1985), xxi.

Figure 14 Have you any women folk worth defending?

image forms the backdrop of the poster and may be a view of the port of Dun Laoghaire, 8 miles south of Dublin city.

Although during the First World War women did occupy spaces normally inhabited by men, the use of female imagery in more conventional roles persisted. In Ireland the presentation of the nation as female has a long history. The woman in visual imagery was employed paradoxically and at different moments in defence of empire and in nationalist discourse.[108] Hibernia in this context asks her men

[108] N. C. Johnson, 'Sculpting heroic histories: celebrating the centenary of the 1798 rebellion in Ireland', *Transactions of the Institute of British Geographers*, 19 (1994), 78–93; Johnson, 'Cast in stone: monuments, geography and nationalism', *Environment and Planning D: Society and Space*, 13 (1995), 51–65 ; C. Nash, 'Renaming and remapping', *Feminist Review*, 44 (1993), 39–57; Nash,

Figure 15 Will you answer the call?

to defend the gendered landscape of Ireland. In Figure 16 orthodox gender roles are again employed to shame men to enlist. The domestic sphere represented and occupied by mother, child and grandfather represents the fundamental, intimate space of private family relationships. But this space is the microcosm of the larger society, made up of millions of other similar family spaces which men must defend. Although war may revolve around complex geopolitical and economic issues, for the ordinary person war can be mediated effectively as literally a defence of one's

'Embodying the nation: the west of Ireland landscape and Irish identity' in B. O'Connor and M. Cronin, eds., *Tourism in Ireland: a critical analysis* (Cork, 1993), 86–112; Nash, 'Remapping the body/land: new cartographies of identity, gender and landscape in Ireland' in A. Blunt and G. Rose, eds., *Writing women and space: colonial and postcolonial geographies* (New York, 1994), 227–50.

Figure 16 Is your home worth fighting for?

house and home. The ideological and political issues at stake can be simplified for popular consumption, leaving the larger questions for political leaders and military strategists.

Not all recruitment posters adopted conventional gender tropes. The moral principle of protecting the small, neutral, independent nation-state of Belgium from German atrocity could be combined in interesting ways with a gendered willingness to serve. Figure 17 depicts an assertive, active woman, neither allegorical nor angelic, imploring and challenging her men-folk to serve, to exercise moral authority in the defence of a blazing Belgium. The dominant female image, in her tight-bodiced dress and rifle in hand, seeks to remind the reader of the willingness of women to meet the moral imperative to serve, and to make the necessary sacrifice.

Figure 17 For the glory of Ireland

Her moral authority is at once universal and simultaneously nationalised – 'for the glory of Ireland'. The family group in the background are walking towards the shore, perhaps bidding farewell to Belgium, as many Irish families had bid farewell to their loved ones boarding emigrant ships over the previous sixty years. The man in this poster, again dandily attired as was the man in Figure 12, is challenged at a number of levels. His moral integrity, his masculinity and his nationality are under review. The juxtaposition of a small nation in flames because of the bullying behaviour of its larger neighbour was read, in some quarters, as a counterpoint to Britain's relationship with Ireland. With Home Rule on the statute books and Redmond's appeal to Irish men to play a role in the defence of the rights of small nations, the image in this poster could suggest Britain's willingness to protect its smaller neighbour. This position, however, was hotly contested in separatists' view of the war.

Together these posters provide a flavour of the way the war was mediated in Ireland.[109] If recruits were to enlist to fight a bloody and protracted war, a set of imaginative tropes had to be employed which would muster support in the specific context in which recruitment was taking place. Patriotism had to be encouraged in a variety of ways and the men who volunteered had to find a set of principles which made their sacrifice seem justified. Thus, while each combatant state employed a series of common motifs to boost volunteering, the style and message had to be tailored to local circumstances as universalistic imperatives alone would not encourage young men to enlist. To make sense of the war meant making sense of it in divergent ways that responded to localised political and ideological contexts. In Ireland posters needed primarily to capture the nationalist imagination. For unionists the war did not need to be domesticated on these terms. Appeals to king, country and empire were sufficient stimuli to serve and the trope of duty to a British cause could be read far more unproblematically than for those with nationalist leanings. Hence the deployment of images and the use of text which recognised the relationship between Ireland and Britain as articulated by constitutional nationalists.

Although posters convey the grand narrative of recruitment and the myth of soldiering, the precise motives lying behind enlistment can never be fully identified. Individuals will provide their own reasons for engagement, many of which will reflect the broader themes outlined here. The nationalist MP and scholar Tom Kettle, for instance, claimed the justness of the cause was his motive to enlist. For others, economic factors may have been more important. Derry man Jim Donaghy said he enlisted out of necessity after he was laid off from work, while James English, a Wexford labourer, calculated that his family would be 154 per cent better off if he joined the forces. Paradoxically, recruitment was often most intense in cities with large industries that had well-organised and stable workforces. In Belfast, for instance, the men most likely to enlist were drawn from well-paid jobs in the shipbuilding industry. A sense of communal identity and, in this case, a strong unionist identity, combined with the repartee of workmates, motivated many men.[110] Similarly, a spirit of adventure and excitement generated through the recruitment campaign also animated decisions to volunteer. Wallace Lyon, a

---

[109] Of the 203 posters in this collection, 157 had some Irish reference and although it is difficult to categorise the posters in terms of content, it has been suggested that 83 appealed to some sense of Irish patriotism, while 43 focused on duty and honour. See Tierney, Bowen and Fitzpatrick, 'Recruiting posters', 53. The examples I selected for inclusion in this chapter attempted to draw on these two dominant themes, as well as a number of the general recruitment motifs that made little or no specific reference to any Irish cultural context.

[110] Jeffery, *Ireland and the Great War*.

young Protestant who had worked in India, claimed: 'I had enjoyed pig sticking in India, and I thought it would be great fun to try my hand at the Uhlans.'[111] Tom Barry, the celebrated IRA man from west Cork, said: 'I went to the war for no other reason than that I wanted to see what war was like, to get a gun, to see new countries and to feel a grown man.'[112] Adventure, a sense of masculinity, and the excitement of travelling to far away places all underpin this statement and these sentiments can be found among many young recruits. Although individual motives were various, communicating the spirit of war and influencing public opinion was, in large measure, mediated through the recruitment campaigns. In Ireland this campaign acted as the larger window through which various factions of opinion would subsequently remember and make sense of the war.

Conclusion

Together with other parts of the empire, Ireland presented a recruiting ground for the raising of the new armies required by Kitchener to take the war to the Germans. While initial mobilisation in Ireland was drawn from the standing army, it quickly became evident that additional recruits would be required to fill the three new army divisions. The existence of four 'unofficial' armies in Ireland, set up to tackle the national question, presented both an opportunity and a threat to Britain's war strategy. Both the UVF and the Irish Volunteers mobilised their supporters to engage in the war effort but the absence of one section of volunteers represented a small but significant segment of opposition to the conflict. An assessment of the geography of recruitment indicates a greater willingness among Protestants to answer the call than their Catholic counterparts. Ulster consistently provided the highest ratio of enlistment, with Belfast exceeding Dublin in terms of numbers. Outside of Dublin and the north-east, the areas of heaviest recruitment were found in the midlands stretching from Longford to Tipperary. Weakest levels of enlistment occurred along much of the Atlantic seaboard (with the exception of Sligo). Overall, 43 per cent of all Irish recruits were Protestants.[113] Nonetheless, what is certain is that both Protestant and Catholic did contribute significantly to the war and both served well whether for king, empire or nation.

This chapter has focused on the official recruitment mechanisms organised by the state to recruit in Ireland. Of course, visual appeals to enlist were part of a larger campaign to sell the war in the Home Front. They often played a crucial role in conveying a sense of the war. In a Waterford context, for instance, advertisements were

reinforced by attractive posters which were used to 'literally besiege the city'. Using 'splendidly localised appeals' and of a convenient size for windows, notice boards and shop-fronts they were displayed in every conceivable and conspicuous place. The posters were in great

---

[111] *Ibid.*, 21.   [112] *Ibid.*
[113] See Fitzpatrick, 'Militarism in Ireland'.

public demand and had been fixed to hoardings and dead walls throughout the city as well as on buildings.[114]

The mass publicity given to the recruitment campaign offers us insights into the dominant tropes to emerge in an Irish context. The convergence of universalistic appeals to duty with highly localised frames of reference underline the astuteness of both the producers and consumers of war recruitment propaganda. Using posters as signifying systems operating within a larger metalanguage of war, this chapter reveals how local, national and international considerations informed both the production and the representation of the war to a diverse Irish audience. In the following chapter I shall turn to considering the first acts of remembrance of the war dead through an analysis of parades. While the recruitment of an army in Ireland magnified some of the divisions on the island, the simmering presence of conflict at home, culminating in the rebellion in 1916, would inform subsequent rituals of public remembrance in the post-war period and it is towards this issue that I now wish to move.

---

[114] Quote on the recruitment campaign in Waterford in March 1915 from T. Dooley 'Politics, bands and marketing: army recruitment in Waterford city, 1914–15', *Irish Sword*, 18 (1991), 212.

3

Parading memory: peace day celebrations

Life springs from death; and from the graves of patriot men and women spring living nations[1]

This extract from Patrick Pearse's renowned oration of 1915 at the graveside of the Fenian, Jeremiah O'Donovan Rossa, reminds us of the powerful political and symbolic role of public commemoration in the politics of everyday life in early twentieth-century Ireland. The previous century had provided several important precedents for commemorating the death of political leaders as the funerals of O'Connell, Parnell and MacManus testify. Commemoration, however, was not confined to individual leaders. The politics of memory generated by the centenary celebrations of the 1798 rebellion, represented through the fusion of the heroic priest-leader and the archetypal peasant in public statuary, illustrates that collective memory could also be aroused through the remembrance of an anonymous rebel soldier.[2] As Whelan puts it, in his examination of official and popular readings of the rebellion, 'besides its Catholic-nationalist reading, the centenary was pivotal in knitting together the strands of nationalist opinion which had unravelled in the acrimonious aftermath of the Parnell split'.[3]

Over two decades later, commemorating the dead who served in Irish regiments in the First World War would similarly challenge cultural allegiances in Ireland, both in nationalist and unionist quarters. The peace parades of July 1919 established the initial framework for commemoration. The public spectacle staged in cities and towns around the country in 1919 provides insights into how the war was

---

[1] This quotation comes from Patrick Pearse's graveside panegyric delivered at the funeral of the Fenian leader Jeremiah O'Donovan Rossa in Glasnevin cemetery in 1915. See P. Mac Aonghusa, and L. Ó Réagáin, eds., *The best of Pearse* (Cork, 1967), 134.

[2] For an overview of the centenary celebrations of 1798 and the associated iconography see the following: T. J. O' Keefe, 'The 1898 efforts to celebrate the United Irishmen: the "98 centennial" *Éire-Ireland*, 23 (1988), 51–73; O'Keefe, ' "Who fears to speak of '98": the rhetoric and rituals of the United Irishmen centennial, 1898', *Éire-Ireland*, 28 (1992), 67–91; Johnson, 'Sculpting heroic histories', 78–93.

[3] K. Whelan, *The tree of liberty* (Cork, 1996), 174.

calibrated in the popular imagination at a moment when the Home Rule crisis was not yet resolved and the Easter rebellion of 1916 was fresh in the public's memory. Although the war has been treated by some scholars as a defining juncture in provoking a modern memory, in the Irish case popular interpretations of the conflict cannot be easily disentangled from the pre-war political conditions on the island. For one historian, 'honouring the dead was not simply a matter of paying due respects – it forms a potent element in the endorsement of a particular political culture or the creation of an alternative one'.[4] The mapping of commemorative space in Ireland in 1919 was a controversial exercise from the outset. While all participating states had to face the challenge of confronting the losses endured during the war and dealing with the inadequacy of a romantic view of memory in doing so, in the Irish case the exercise of social memory rubbed up against a whole suite of immediate conflicting allegiances, and these allegiances would find material expression in the spatialisation of a 'national' imaginary.

In this chapter I will focus on how the memory of the dead of the First World War in Ireland was articulated through an analysis of the Peace Day parades of July 1919 and subsequent Armistice Day commemorations. Irish men and women participated in significant numbers in the war, and although there were marked religious and regional patterns to enlistment, focusing on Ulster, Dublin and the midland counties, this chapter contends that the circumstances under which Irish people participated in the war partly explain the significance that was subsequently attached to the war through public commemoration. The parades themselves represent the first attempt in Ireland to attach cultural and political meaning to the war and as such they laid the foundations for the manner in which future generations would make sense of the war. Drawing again from Roland Barthes and his analysis of the role and meaning of public spectacle, this chapter analyses the parades as spectacles where 'what is expected is the intelligible representation of moral situations which are usually private'.[5] While remembering the dead is frequently conceived as a personal affair, commemoration of war dead became a public, collective event, which implicated the society as a whole. Through analysing commemoration as large-scale spectacle, I suggest that collective memory is maintained as much through geographical discourses as historical ones. Spectacle constructs the spatial and temporal limits to popular understandings of the past, and in so doing it underlines how universal principles of bereavement are locally mediated, and this was achieved through the actual patterns of spectacle.

The first part of this chapter positions the examination of commemoration within a larger academic literature on the Great War and the politics of memory. The second part offers a discussion of parades as an exemplar of public spectacle in dramatic form. Using Barthes' analysis of wrestling as a guiding framework,

---

[4] P. Travers, 'Our Fenian Dead: Glasnevin cemetery and the genesis of the Republican funeral' in J. Kelly and U. MacGearailt, eds., *Dublin and Dubliners* (Dublin, 1990), 52.

[5] R. Barthes, 'The world of wrestling' from *Mythologies*, trans. Annette Lavers (New York, 1972), 22.

the chapter then interprets the Peace Day parades of July 1919 as a moment when confused allegiances were brought sharply into focus, and where remembrance of the dead had at once a unifying and disintegrating effect on public consciousness. The final part of this chapter provides an overview of commemorative spectacle in Ireland in the years after independence, where the processes of nation-building in the Irish Free State and the ongoing cultivation of a separate identity in Ulster generated mixed responses to the memory of the First World War.

The politics of memory

Although James Joyce made only one direct reference to the Great War, literary historians have contended that *Ulysses* 'constitutes a response in content and form, not only to World War I, the Easter Rising, and other upheavals, but to the preceding quarter of a century – a period of intensified imperial and national rivalries, of technological innovation, of social change'.[6] Stephen Dedalus, the novel's principal character, makes the complaint that history is the nightmare from which he is seeking to escape.[7] For European society, the years 1914–18 can also be seen as a nightmare out of which it was trying to escape. The release, however, was never complete: fragments of the nightmare persisted in the memory of both the individual soldier and the larger society. The structuring of this post-war memory, both private and public, entails some discussion of the relationship between history as past events, and history as a narrative account of past events. For the historical geographer the written account is central, but as Frederic Jameson points out, the past itself 'is not a text, not a narrative'.[8]

For historians in the nineteenth century the text may have been construed as a straightforward presentation of what actually happened. In this century, it has been more fully acknowledged that the evidence of history cannot be so easily separated from the interpretation built upon it.[9] This is especially true of efforts to situate the First World War in social, economic and intellectual history. For instance, feminist historians have begun to address the impact of the war on gender relations and they have drawn quite diverse conclusions. Some have viewed the war as a deciding moment in the re-articulation of gender roles through documenting the extension of female social, economic and sexual freedoms during the conflict.[10] Others, however, have interpreted the evidence in a different manner. Margaret and Patrice Higonnet have claimed 'to trace the continuity behind the wartime material changes in women's lives. That continuity lies in the subordination of women's new roles

---

[6] J. Fairhall, *James Joyce and the question of history* (Cambridge, 1993), 164.

[7] J. Joyce, *Ulysses*, edited with introduction by Jeri Johnson (Oxford, 1993). Originally published in 1922.

[8] F. Jameson, *The political unconscious* (Ithaca, 1981), 35.

[9] For a further discussion of this point see Robin Collingwood, *The idea of history* (Oxford, 1946).

[10] See S. M. Gilbert and S. Gubar, *No man's land: The place of the woman writer in the twentieth century. Vol. 2: Sexchanges* (New Haven, 1989).

to those of men, in their symbolic function, and more generally in the integrative ideology through which their work is perceived.'[11] This example illustrates that our account of past events cannot rely on the robustness of the evidence alone; it is also dependent on the theoretical framework guiding it.

Representations of the war and the construction of a collective memory of the conflict have also been subject to diverse analyses. Literary historians have argued that the war represented a critical juncture in the evolution of an ironic modernism, particularly expressed in the visual arts and literature.[12] Together these studies have focused attention on elite responses to the war. Alternative views of commemoration stress the linkages between post-war memory and the cultivation of nationalist politics, especially in Germany and Italy.[13] One historian claims that 'Modern memory was born not just from the sense of a break with the past, but from an intense awareness of the conflicting representations of the past and the effort of each group to make its version the basis of national identity.'[14] A number of studies have stressed the need for a contextual approach to commemoration that integrates into the analysis the voices of a variety of different actors: soldiers, veterans' organisations, the public and the state.[15] Geographers too have examined landscapes of war and memory where they have stressed the debates underpinning the commemoration of war dead and the construction of national or regional identities.[16]

While the distinction between modern and traditional memory was identified in Chapter 1 and some of the shortcomings of the dualism outlined, the distinction between elite and popular memory yields some interesting trends historically. Although elite, archival memory colonised time and partitioned territory into bounded space, popular memory was more episodic, non-linear and sometimes more local. Time was not measured from single beginnings but from centres where time could move backwards and forwards through living memory.[17] The gradual transformation of popular memory from a living one to an archival one began with the

---

[11] Higonnet and Higonnet, 'The double helix', 39.

[12] This view is most cogently argued by Fussell, *The Great War and modern memory*. It is also supported by E. Leed in his study of the psychological impact of the war on men *No man's land: combat and identity in World War One* (Cambridge, 1979); and in the examination of the impact of the war on English culture by Hynes, *A war imagined*.

[13] See G. Mossé, *Fallen soldiers: shaping the memory of two world wars* (Oxford, 1990).

[14] J. R. Gillis, 'Memory and identity: The history of a relationship' in R. Gillis ed., *Commemorations: the politics of national identity* (Princeton, NJ, 1994), 8.

[15] Recent work includes Gregory, *The silence of memory*; R. W. Whalen, *Bitter wounds: German victims of the Great War* (Ithaca, 1984); A. Becker, *Les monuments aux morts: mémoire de la Grande Guerre* (Paris, 1988); Vance, *Death so noble*.

[16] Heffernan, 'For ever England: the Western Front and the politics of remembrance in Britain'. In terms of the American civil war see Winberry 'Lest we forget', 107–21. For France see H. Clout, *Mémoires de Pierre: les monuments aux morts de la Première Guerre Mondiale dans le Pas-de-Calais* (Calais: 1992); *After the ruins: restoring the countryside of Northern France after the Great War* (Exeter, 1996).

[17] Gillis, 'Memory and identity'.

political and economic transformations of the late eighteenth century. What had once been the preserve of elites such as the Church, the monarchy and the aristocracy, gradually became democratised as the demand for commemoration spread to the urban middle classes and the working classes. While this process varied from place to place, popular memory was re-born to fashion a new future as well as articulate a shared past. In Europe, for instance, the cult of new beginnings found one of its earliest expressions in the celebrations of the French Revolution and the installation of a national holiday on 14 July, marking the Fall of the Bastille.[18] The new French republic's extraordinary effort to alter time consciousness is reflected in its declaration of 1792 as Year I of its calendar, a potent symbol of literally new beginnings.[19] The key facet of this change towards the popularisation of memory is that 'it relies entirely on the materiality of the trace, the immediacy of the recording, the visibility of the image'.[20] In terms of the First World War this is particularly relevant as Europeans adopted the American model of a military cemetery where officers and men would be interred side by side, and where memorials both in style and size were different from those which went before. Remembrance was materialised through a wide variety of acts of commemoration and whilst the industrialisation of warfare had produced enormous casualties, the industry of remembrance produced a range of memorialising strategies, some of which were modernist but many of which were also romantic and traditional.

In terms of the concerns of this chapter several issues emerge. First, until recently, much of the discussion dealing with the First World War did not deal with the ways in which the war was interpreted by more minor actors in the conflict, whose relationships with the bigger powers (even as allies) were complex and contested.[21] Second, a focus on the traditional/modern debate in positioning the war in cultural history, tends to overdichotomise processes of change. While these labels may be useful heuristic devices for academic historians to structure their analyses, the coexistence of competing forms of popular remembrance and representation, in time and in space, seems critical for understanding the conflict and for positioning intellectually our idea of memory. This is related to a third element in the historiographical debate. While a contextualised approach to historiography is frequently propounded, the geographies of remembrance are generally subsumed by the histories of memory in ways which treat space as epiphenomenal to historical

---

[18] M. Ozouf, *Festivals and the French Revolution* (Cambridge, MA, 1988).

[19] E. Zerubavel, *Hidden rhythms: schedules and calendars in social life* (Chicago, 1981).

[20] Nora, 'Between memory and history', 13.

[21] Over the past decade there have been an increasing number of studies dealing with the 'minor' actors in the Great War. See, for instance, A. Gaffney, *Aftermath: remembering the Great War in Wales* (Cardiff, 1998); Inglis, *Sacred places*; A. Thomson, 'The Anzac myth: exploring national myth and memory in Australia' in R. Samuel and P. Thomson, eds., *The myths we live by* (London, 1990), 73–82; J. Pierce, 'Constructing memory: the Vimy memorial', *Canadian Military History*, 1 (1992), 1–3; Osborne, 'Warscapes, landscapes, inscapes' in Black and Butlin, eds., *Place, culture and identity*, 311–333; P. Baker, *King and country call: New Zealanders, conscription and the Great War* (Auckland, 1988); M. McKinnion, *New Zealand historic atlas* (Auckland, 1997).

process. Consequently, the sites of commemorative activity tend to be treated as reflective of the meaning attached to the war rather than constitutive in the creation of that meaning. By focusing on a comparatively peripheral participant, Ireland, and by taking seriously the public spectacle and the drama involved in remembrance, this chapter seeks to overcome some of these difficulties.

Analyses of Ireland's participation in the Great War, both north and south of the border, amount to little more than a handful of books. Some concentrate on the military history of a specific division and its role in particular battles;[22] others are records of the memoirs of individual soldiers.[23] Recently there has been a growth of interest by academic historians in documenting Ireland's efforts to commemorate the war.[24] Despite the importance of the 36th Ulster Division in Northern Ireland and its role in popular understandings of the past, especially among the unionist population, academic analysis is still comparatively slight. Most recent commentators attribute the lack of a comprehensive historiography of the war to a nationalist political agenda by Irish historians. This may account for the absence of a substantial body of research in the Irish republic, but it does not account for a similar absence in Northern Ireland. A more deciding factor may relate to the practice of historiography in Ireland. Until recently there has been an overwhelming emphasis on the political history of the island especially for the period leading to independence and partition. This may have diverted attention away from the Great War except as a contextual backdrop to political events at home. The emergence of economic and social history, however, has broadened the remit of academic studies in Ireland. The blurring of boundaries between disciplines has also contributed to an emerging emphasis on cultural approaches to the past, which combine the work of literary critics, philosophers, historians, sociologists and geographers.[25] Together these changes have spurred a renewed interest in the war and have shifted emphasis away from the narrower concerns of regimental histories to broader themes related to representation.

[22] The following are conventional military histories of specific regiments: T. Denman, *Ireland's unknown soldiers: the 16th (Irish) Division in Great War* (Dublin, 1992); B. Cooper, *The tenth (Irish) division in Gallipoli* (Dublin, 1993); Johnstone, *Orange, green and khaki.*

[23] For accounts based on the memoirs and oral histories of Irish participants in the Great War see P. Orr, *The road to the Somme* (Belfast, 1987); Dungan, *Distant drums*; Dungan, *Irish voices from the Great War.*

[24] The work of academic historians includes G. Boyce, *The sure confusing drum: Ireland and the first world war* (Swansea, 1993); Boyce, 'Ireland and the First World War', *History Ireland*, 2 (1994), 48–53; Fitzpatrick, *Ireland and the First World War*; T. Bartlett and K. Jeffery, eds., *A military history of Ireland* (Cambridge, 1996); K. Jeffery, ed., *Men, women and war* (Dublin, 1993); Jeffery, 'Irish artists and the First World War', *History Ireland*, 1 (1993), 42–5; Jeffery, 'Irish culture and the Great War', *Bullán*, 1 (1994), 87–96; Jeffery, *Ireland and the Great War*; J. Leonard, 'The twinge of memory: Armistice Day and Remembrance Sunday in Dublin since 1919' in English and Walker, eds., *Unionism in modern Ireland*, 99–114.

[25] This includes the work of Whelan, *The tree of liberty*; D. Kiberd, *Inventing Ireland* (London, 1995); L. Gibbons, *Transformations in Irish culture* (Cork, 1996); D. Lloyd, *Anomalous states: Irish writing and the postcolonial moment* (Dublin, 1993).

The spectacle of memory

Unlike formal academic histories, where an account of the past is conventionally structured around the concatenation of episodes into a narrative, public memory may be more suitably articulated as a spatial arrangement of objects around a spectacle. The Dutch historian Leersen puts it as follows: 'one way of unifying history [is] to rearrange its consecutive events from a narrative order into a spectacle, a conspectus of juxtaposed "freeze-frame" images'.[26] The memory of four consecutive years of war, for instance, can be foreshortened into a single commemorative event. The collapsing of time into space through the annual rehearsal and repetition of a spectacle provides a framework, not only for understanding remembrance, but also for the public enactment of forgetfulness. Drawing on Debord's *Society of the Spectacle*, geographers have begun to theorise the extent to which spectacle has become the total lens through which modern society is experienced and controlled. Ley and Olds suggest that 'spectacle is the manifestation of the power of commodity relations, and the instrument of hegemonic consciousness',[27] where the masses of spectators are rendered passive and duplicitous in their own impotence. This view of spectacle has recently been modified and the monolithic control of the spectator by those creating the spectacle has been challenged through analysing parody and other subversive uses of spectacle.[28]

The genealogy of the spectacle metaphor and the different meanings associated with the term has been explored. These range from spectacle as ordinary display to spectacle as 'the sense of a mirror through which truth which cannot be stated directly may be seen reflected and perhaps distorted'.[29] This latter view of spectacle derives from Barthes' fascinating work on the ways in which spectacle works. In his discussion of the 'spectacle of excess' witnessed in popular wrestling, and drawing parallels with ancient theatre, he claims that 'What is thus displayed for the public is the great spectacle of Suffering, Defeat, and Justice.'[30] Analysing the cultural meaning of spectacle, Barthes has stressed the significance not only of words and actions but also of objects themselves (the bodies of the wrestlers) as symbols in the production of meaning.[31] For Barthes these bodies of the wrestlers and the ritualised behaviour in which they engage come to represent the moral drama of modern society.

The strength of this approach to the study of remembrance of the Great War is that it was popularly represented precisely through large-scale drama or theatrical performance. The construction of a spectacle of remembrance translated individual

---

[26] J. Leersen, *Remembrance and imagination: patterns in the historical and literary representation of Ireland in the nineteenth century* (Cork, 1996), 7.

[27] D. Ley and K. Olds, 'Landscape as spectacle: world's fairs and the culture of heroic consumption', *Environment and Planning D: Society and Space*, 6 (1988), 194.

[28] For a fuller discussion of this critique see A. Bonnett, 'Situationism, geography and poststructuralism', *Environment and Planning D: Society and Space*, 7 (1989), 131–46.

[29] Daniels and Cosgrove, 'Spectacle and text', 58.   [30] Barthes, 'The world of wrestling', 23.

[31] Barthes, *The elements of semiology*.

responses to loss and victory into a collective response, where the relationship between the 'actors' in the spectacle, the audience viewing it, and the geographical setting which framed it, all created the context for interpretation. In his discussion of wrestling, Barthes stressed these precise types of connections. The exaggerated antics of the wrestlers, the moral expectations of the audience and the arenas in which the meaning was adjudicated were all interrelated. For Barthes wrestling was not a sport, viewed to see who would win or lose on the basis of physique alone; it was a spectacle where the ethics of the physical encounter were negotiated.

While modern-day wrestling may seem a far cry from the slaughter of the First World War, the question of the intelligibility of death, and in this case the prodigious loss of life, is germane. Each death was simultaneously a private moral matter (for family and for friends) and a public one (for states and for armies). The response of a civilian audience to that which they themselves did not experience directly raised questions about the moral and political meaning of modern warfare. European society, in the aftermath of the war, attempted to present and reconcile these questions through staging annual parades and creating commemorative landscapes. By treating these as ritual spectacles, albeit considerably different in kind from more orthodox spectacular events, we begin to unravel the ways in which large-scale death could be culturally and morally harmonised in a peacetime environment. What Barthes offers us is a way of grasping the moral universe in which a spectacle is staged and the significance of the staging itself as the arena in which questions of suffering and bereavement can cohere with issues of justice and rectitude.

An account of the past relayed through public spectacle, like narrative history, is partly mediated through the lens of current political preoccupations. In the case of Ireland this involved constructing a commemorative spectacle when the pre-1914 divisions were not eliminated, the constitutional position of Ireland within the union was unknown and the Easter rebellion was still fresh in the public mind. These facts add a specific dimension to Ireland's acts of remembrance that differentiate it in important ways from the fashioning of memory in Britain and France. The manner in which the spectacle was produced and received across Ireland varied considerably.

The spectacle of remembrance: Peace Day 19th July 1919

In the aftermath of the Great War, rituals to mark its end and to commemorate its dead were quickly under way. While 19 July 1919 was designated 'Peace Day' in Britain and marked in London by the parading of 18,000 troops past the Cenotaph in Whitehall,[32] plans were made also in Ireland to mark this day. In Dublin, by early 1919, there were proposals afoot to establish an Irish national war memorial. A committee for that purpose, headed by Lord French, the lord lieutenant for Ireland, was established. The initial intention of the committee was to erect a

---

[32] Hynes, *A war imagined.*

war memorial home for ex-servicemen visiting or passing through Dublin and to establish a record room, which would contain the parchment rolls of all fallen Irish soldiers.[33] Although Lord French hoped to have the plan under way by the end of 1918, he realised that 'Nothing, however, could be done until the whole of loyal Ireland was brought into council.'[34] While the committee, by and large, supported the proposal to locate the memorial home in Dublin, there was a strong view that it ought to be a 'symbol of unity' on the island, uniting north and south, Catholic and Protestant.[35] Although supporting the proposed national memorial, representatives from the north of Ireland made it known that they also had their own plans. The mayor of Belfast observed that the Church of Ireland and the Presbyterian Church had already begun erecting commemorative plaques in their churches. The mayor of Derry confirmed that his city would fund a memorial to honour their dead. Although Ulster may have been perceived to be pursuing a more independent route, Captain Dixon MP reassured the public that Ulster's loyalty to the crown did not undermine their view that the soldier from Clare (west of Ireland) was equal to the soldier from Shankill (west Belfast) and should be remembered as such.[36]

Dublin had experienced a variety of large public spectacles and parades in the second half of the nineteenth century and thus was accustomed to staging large-scale processions. Most notable amongst these were the public funerals of prominent nationalists including Daniel O'Connell, Terence Bellew MacManus and Charles Stewart Parnell. All were buried in Glasnevin cemetery, in the city's north side. In O'Connell's case his body lay in state at the Catholic Pro-Cathedral in the north inner city, before being taken on a circuitous route to the cemetery. The route proceeded from the north inner city along Sackville Street to the south inner city before returning north again to the graveyard.[37] In the case of MacManus and Parnell, the processions also included the north and south inner city. In the routing of these funeral corteges Travers reminds us that they were 'primarily designed to evoke the memories of dead patriots'.[38] The unveiling of the foundation stone of the O'Connell monument on Sackville Street in 1864 involved a procession from Merrion Square to the site. The final unveiling of the statue in 1882 also revolved around a procession 'that took participants past a range of buildings with which O'Connell had some form of association'.[39] These included buildings north and south of the river. Similarly, royal visits in the nineteenth century were marked by public performances including visits to the Viceregal Lodge, the Georgian mansion in the Phoenix Park (and now official home of the Irish president).[40]

[33] *Irish Times*, 4 July 1919.   [34] *Irish Times*, 18 July 1919.

[35] The Provost, Trinity College Dublin, expressed this view most forcefully, *Irish Times*, 18 July 1919.

[36] *Irish Times*, 18 July 1919.

[37] For a detailed discussion of the routes of these funerals see Travers, 'Our fenian dead', 52–72.

[38] *Ibid.*, 58.

[39] Y. Whelan, 'Monuments, power and contested space – the iconography of Sackville Street (O'Connell Street) before Independence (1922)', *Irish Geography*, 34 (2001), 25.

[40] Y. Whelan, 'The construction and destruction of a colonial landscape: commemorating British monarchs in Dublin before and after Independence', *Journal of Historical Geography* (forthcoming).

The celebration of Peace Day in Dublin took on the characteristics of a spectacle. By royal proclamation the day was declared a bank holiday and this was observed by many of the mercantile community in the city. The victory parade was a well-organised public event with the route of the march published in the national press. The parade began at Dublin Castle, the centre of 'colonial' administration in Ireland. The participants assembled in the lower Castle yard between 9.30 and 10.30a.m., and here the order of procession was organised to begin at 11.30a.m. The sequence consisted of a leading troop of mounted police, followed by the Irish Guards Piper's Band, transported from Windsor for the event. Demobilised Irish soldiers and sailors followed, marshalled according to regiment, and led by their own officers. As many as possible were clothed in khaki. Following the troops was the commanding officer and his staff. Different units of artillery and cavalry were next in line and were followed by representatives of the RAF, WRAF, WAAC, Red Cross and VADs. Bringing up the rear was a huge display of tanks and armoured cars. The procession comprised about 20,000 people, of which 5,000 were demobilised soldiers and sailors.[41] Yet not all veterans' organisations participated. The Discharged Soldiers and Sailors Federation was not represented and 2,000 to 3,000 Irish Nationalist Veterans boycotted the event.[42]

The parade followed a designated route along the thoroughfares of the south inner city terminating at St Stephen's Green (Figure 18).[43] Notably the parade was not routed along Sackville Street[44] (the main street of the city and the nexus of the Easter rebellion), partly because many of the buildings along the street were still under reconstruction since the rebellion. The focal point of the procession was at the Bank of Ireland, College Green, where a stand for the viceregal party had been erected the previous day. The irony of this space did not go unnoticed. The editorial of the *Freeman's Journal* observed: 'By a refinement of irony in keeping with the best traditions of Dublin Castle, the Viceroy and Chief Secretary elected to take the salute in front of the old Parliament House, emphasising the fact that what counts in Ireland is not the will of its people ... but the power of its rulers to mass bayonets, tanks and field-guns.'[45] The playing of the national anthem and the hoisting of the Union Jack greeted the arrival of the lord lieutenant to College Green. The soldiers took the salute here. This space acted as the symbolic keystone of the parade. Opposite the Bank, in the forecourt of Trinity College,

---

[41] *Irish Independent*, 21 July 1919.   [42] *Freeman's Journal*, 21 July 1919.

[43] Based on reports published in *Irish Times*, 21 July 1919.

[44] Sackville Street formed the principal north–south artery of the city. Begun in the 1740s by the prosperous city banker, Luke Gardiner, it was originally intended as an elongated residential square (called the Mall). It was transformed in 1784 by the building of Lower Sackville Street by the Wide Street Commissioners. The latter were also responsible for the construction of Carlisle Bridge linking Sackville Street with the newly widened south city streets of D'Olier and Westmoreland. M. Craig, *Dublin 1660–1860* (Dublin, 1980). By 1900 the hub of Dublin's tramway system was at Nelson's Pillar in the middle of Sackville Street, opposite the General Post Office (GPO). Built between 1814–1818 the GPO had been extensively modernised and re-opened to the public in March 1916. One month later the entire building except the façade lay in rubble after the Easter Rising.

[45] *Freeman's Journal*, 21 July 1919.

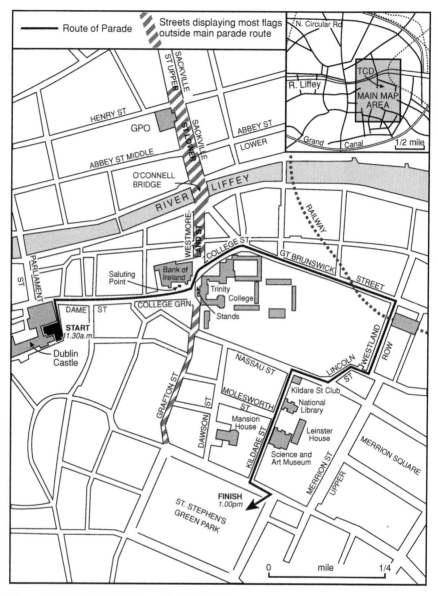

Figure 18 Route of Dublin's parade, July 1919

two stands were occupied by wounded veterans, offering them a vantage point
from which to view the parade and be viewed by the spectators. This junction
along the route provided prized space for spectators to assemble where they could
simultaneously glimpse the procession of military personnel and Britain's state

representative in Ireland. In rather hyperbolic terms the *Irish Times* recorded events as follows:

Politics, dissension, everything are forgotten as Ireland's Viceroy and the Empire's first defender takes his stand under his well-served flag; and for some minutes, at any rate, one felt that every voice in Ireland was paying throaty tribute in honest thanksgiving to a man in whose person the spirit of victory and peace was symbolised.[46]

If the meaning of the war was to be mediated through spectacle, the viceroy, Union Jack and National Anthem provided the necessary symbols of legitimacy. The lord mayor of Dublin and the city's Corporation (of a nationalist political persuasion), however, did not endorse the parade and attendance was left to the discretion of individual council members.[47]

The parade proceeded along the streets skirting Trinity College, and south to St Stephen's Green. A particularly enthusiastic welcome was noted outside the Kildare Street Club,[48] one of the leading social clubs for politicians founded in the eighteenth century during Grattan's Parliament. Indeed Kildare Street represented the aristocratic heartland of the city, boasting the residence of ten peers of the realm over the previous hundred years. The marching of soldiers in clean, well-pressed uniforms, although contrasting with the filth of the trenches, conveyed a sense of orderliness and rationality to the war and mirrored some of the images of soldiering found in recruitment posters in Chapter 2. The parading of the disabled bodies of some soldiers, however, reminded the public of the suffering necessary to achieve the moral and political goal of victory and Barthes has noted that 'Suffering which appeared without intelligible cause would not be understood.'[49] Intelligibility in this instance was conveyed through flying the Union Jack around the city (including at the General Post Office). These flags represented a symbol of the unity within Britain's empire in defending against the forces of German barbarity. Although flags and bunting were most heavily concentrated along the streets of the parade, Grafton Street and Sackville Street were also heavily adorned (Figure 18). At the GPO, the Union Jack, the American Stars and Stripes and the Italian flag were all hoisted to remind the public of the international effort involved in the achievement of victory. The bells of Christ Church Cathedral rang a continuous peal finishing with a volley firing. According to the *Irish Times*, 'the day's events had shown that Dublin was proud to share with the rest of the empire in celebrating the dawn of peace after an anxious vigil'.[50] Not only were the men of various Irish regiments represented, especially the Dublin Fusiliers, but the role of women in the war effort was also acknowledged. Detachments from the VADs, Red Cross and Women's Legion took part in the event,[51] thereby indicating that the war was not solely the preserve of men but necessitated the supportive role of women to ameliorate the suffering. After the parade there was a formidable display of armoured cars and

[46] *Irish Times*, 21 July 1919.   [47] *Freeman's Journal*, 19 July 1919.
[48] *Freeman's Journal*, 21 July 1919.   [49] Barthes, 'The world of wrestling', 23.
[50] *Irish Times*, 21 July 1919.   [51] *Ibid.*

tanks, the first time Dubliners had seen the machinery of war on such a massive scale since the rebellion three years earlier. The conjunction of soldier, nurse, flag and weapon provided the rationale for remembrance and the context for extracting meaning from this drama.

Unlike Barthes' wrestlers, where suffering is staged literally before the eyes of the audience through stylised gestures of pain and passion,[52] the suffering in First World War processions had already been experienced by the soldiers, the wounded and the bereaved. The spectacle thus sought to ameliorate and render comprehensible suffering already endured, rather than to reenact the pain once again. Thus the evidence of battle – the uniformed ranks, the military bands and the weaponry – reminded the audience of the *potential* pain embedded in the amassed armoury, but, at that moment, they were representations of peace or the absence of physical suffering. The orchestrated assemblage of the machinery of war acted then as the neutral symbol of the means to maintain the moral order. Ironically, they became a synecdoche of civility and righteousness rather than symbols of death and destruction. But they did not stand for such values in isolation. Their moral weight was made meaningful through the iconography surrounding them: the flags of empire; the viceregal entourage; the government's buildings; the houses of learning; the peacetime conditions of the streets. The route of the march was not just a material backdrop, the cover within which the real tale was told or read; it was an intrinsic part of the tale itself. Indeed, it configured the spectacle in particular ways in the hope that it would be interpreted uniformly.

The audience, however, proved to be discriminating in its celebration of peace. While crowds assembled along the route of the parade, enthusiasm was muted in places. For instance, on Brunswick Street and Westland Row, 'Some cheers were raised as the demobilised soldiers passed but the regular troops were received for the most part in silence.'[53] This part of the city, important in nationalist circles as one of the centres of the separatists' anti-recruitment campaign and close to the ideological centre of the Rising, was a different symbolic space to Dublin Castle. The soldier in this spectacle was, then, an ambivalent figure. While those who had served in the field of battle could be honoured as a representation of a just cause, 'an externalised image of torture' which the spectator experiences as 'the perfection of the iconography',[54] for the regular soldier in the parade his role in Ireland in the summer of 1919 could not be easily separated from the prevailing debate about Ireland's place in the union. The concept of justice embodied in the figure of the war veteran could not be transferred with ease to the regular soldier. While young girls could carry banners bearing the inscription 'Welcome home' to demobilised soldiers,[55] the moral position of the professional soldier parading the streets of Dublin remained equivocal. For areas in the city with more unionist leanings the soldier could be seen as both reassuring and honourable while in nationalist areas such readings were difficult.

[52] Barthes, 'The world of wrestling'.   [53] *Irish Independent*, 21 July 1919.
[54] Barthes, 'The world of wrestling', 25.   [55] *Evening Herald*, 19 July 1919.

Despite evidence of general support for the parade during daylight, the evening witnessed a number of incidents that challenged the effectiveness of the spectacle. Around 9.00p.m. two soldiers were attacked on their way back to their barracks. Amidst the scuffle that broke out along the city's quays a police sergeant was shot, although the soldiers themselves made it safely back to their barracks. During the evening crowds of Sinn Féin supporters gathered in various parts of the city, particularly around the General Post Office (GPO), brandishing Sinn Féin flags and singing republican songs.[56] While successful peace entertainment was held in private space at military barracks in the city, soldiers found themselves more vulnerable when they entered public space after dark. Dublin thus could launch a large-scale spectacle but there were no guarantees that all the city's citizens would share in it. The pre-war tensions, which animated the recruitment campaign, resurfaced in this commemorative spectacle having been fuelled further by the events of 1916.

Celebrating peace around the country

Although many Irish towns hosted some event of remembrance for the ending of the war, the local political context had a substantial effect on the nature of support. In the city of Cork, Sinn Féin boycotted the celebrations; no flags flew from City Hall nor did Cork Corporation take part in the event. At the city's workhouse Sinn Féin hoisted black flags over the entrance to the building. Similarly, at their own headquarters blinds were drawn and black flags flew. The iconography of death could be used for diverse political ends. Nevertheless, large crowds took part in the parade, but there was serious rioting in the city in the evening.[57] A policeman was shot; soldiers were attacked and Sinn Féin women physically removed blue, white and red badges from the female friends of soldiers.[58] If during the war women placed white feathers on un-enlisted men, in Ireland brandishing symbols of support for the war was at times interpreted as brandishing icons of betrayal. From press reports, parades west of the Shannon were more muted and most businesses remained open. Remote from the administrative centre and from centres of intense recruitment the impetus to create a spectacle of remembrance was much weaker.

In smaller towns in Leinster and Munster, the spectacle of remembrance was greeted with some ambivalence. In Dundalk, most commercial enterprises did not observe the bank holiday. At the Courthouse graffiti read 'Peace now. This world is safe for hypocrisy.'[59] But not all protests in Ireland emanated from nationalist quarters. In Clonmel, the local branch of the Soldiers and Sailors Federation did not take part in the parade in protest at the government's treatment of ex-servicemen.[60] In Tipperary town a Union Jack floated from the General Post Office but a few yards away a republican flag was suspended on telegraph wires spanning the main street. At 11.00a.m. a party of military police, armed with rifles, removed the

---

[56] *Irish Times*, 21 July 1919.   [57] *Freeman's Journal*, 21 July 1919.
[58] *Cork Examiner*, 22 July 1919.   [59] *Irish Times*, 21 July 1919.   [60] *Ibid.*

republican flag to some ironic cheering from the crowd. At a meeting of the Local Government Board of Guardians, a Mr Quinlan noted that 'the Sinn Féin flag ought to have first place in the town'.[61] At a peace dinner that evening mixed feelings were expressed but Monsignor Ryan, a chaplain in Flanders, struck a conciliatory note in his speech. He claimed that Irish men had fought as 'God's soldiers' against the Germans.[62] If the local political battle could not be resolved, an appeal to a wider moral context might be persuasive.

Wexford town similarly represents an interesting case of how an *account* of past events was writ large on the landscape of commemoration. Although one of the principal sites of the 1798 rebellion and its centenary celebrations one hundred years later, Wexford represented a place where constitutional and republican loyalties competed for support. Despite being the birthplace of John Redmond (leader of the Irish Parliamentary Party) and his brother Major Willie Redmond, killed in the war, the holiday was observed only by government offices, banks and foundries. There was a marked absence of public decorations and the only flag hoisted in the parade was the Irish flag. No public body in the town officially took part in the parade. Official forgetfulness can be as potent a gesture as remembrance. Around 500 ex-servicemen, nevertheless, took part in the procession that congregated for speeches in Wexford Park. In his address, town councillor James McMahon argued that although Irish people had gone to war of their own accord, 'the free gift of a free people to fight for freedom. Now the fight was over, they [the people] should seek freedom for themselves and by constitutional agitation secure self-government for Ireland.'[63] He distanced Wexford from the republicanism of Sinn Féin and he condemned their flag as one sullied by crime and shame. He urged the audience to follow constitutional avenues towards Home Rule and to work under the old green flag of Ireland. The Peace Day commemorations in Wexford provided a forum for the national question to be aired and it illustrates how the desire to accommodate the war within a nationalist constitutional agenda continued to find expression in some areas. While remembrance of the dead and the celebration of peace can appear to have universal resonance, the geography of reception underscores the contingency of public support for such events. Both the micro-geography of the routes of parades in specific towns and cities and the regional mapping of political allegiances disclose the difficulty of creating an agreed collective memory, and this was nowhere more transparent than in Ulster.

Ulster remembers: Peace Day August 1919

In Ulster the spectacle of parades has a long genealogy and thus the development of remembrance rituals for soldiers killed in the First World War extended that practice of public commemoration for military victory. In particular, the success of William of Orange at the Battle of Boyne and his accession to the crown provided

[61] *Ibid.*   [62] *Ibid.*   [63] *Freeman's Journal*, 21 July 1919.

the centrepiece of Ulster Protestant commemorative practices. From the eighteenth century onward, Orange parades have been at the heart of the commemorative calendar of Protestant identity. The central role of the 36th Ulster division in the Battle of the Somme anchored the memory of the war around that single battle. In the first two days of the offensive the Ulster Division lost 5,500 (killed, missing or wounded) from a total of 15,000 soldiers. The fact that the first day of the battle – 1 July – coincided precisely with the Battle of the Boyne in 1690 was recognised at the time. The commanding officer of the 36th Ulster Division, on the eve of the offensive, wrote: 'We could hardly have a date better calculated to inspire national traditions amongst our men of the North.'[64] The losses of the first days of the Somme focused Ulster minds on the personal bereavement experienced by close-knit communities. It also cemented a sense of the social nature of Ulster's sacrifices in the war. The Battle of the Somme became the archetype of Ulster's loyalty and defence of the crown.

Although many thousands of other Irish soldiers lost their lives during the course of the war the intensity and catastrophic strategy associated with this battle was particularly acute in the formation of Ulster's collective memory of the war. Indeed in 1916, for the first time in its history, the Orange Order cancelled its annual 12 July parades and observed a 5 minute silence for those killed. The temporal proximity of the Somme to the 12th of July helped to calibrate the war in Ulster memory along a historical trajectory which emphasised Ulster's continued sacrifice for a greater British cause. As Jarman has noted, 'Opposition to Home Rule was no longer couched solely in references to seventeenth-century battles or in abstract politico-religious ideals; it was securely anchored in the events of the recent past.'[65] The First World War and the Somme in particular would be incorporated into the practices of social memory in Ulster in the years immediately after the war and in what would become Northern Ireland.

In terms of the July peace day celebrations the *Belfast Newsletter* claimed that 'local considerations' (the celebrations associated with 12 July Orange parades) would mean the postponement of the civic celebrations until 9 August. This date would also allow the viceroy to take the salute in Belfast and honour Ulster's contribution to the war effort in an event separate from his role in the Dublin parade. The city did observe the day with an official pageant of military personnel which consisted exclusively of English and Scots regiments. More significantly, in the Orange parades of 1919 the connections between the Somme and the Boyne were first displayed. The Hydepark Loyal Orange Lodge 1067 unveiled their banner which portrayed King William on one side and the Battle of the Somme on the other.[66] The inclusion of the Somme in the iconography of the Orange Order extended the sightlines of Protestant social memory while simultaneously narrowing

---

[64] Quoted by Jeffery, *Ireland and the Great War*, 56.

[65] N. Jarman, *Material conflicts: parades and visual displays in Northern Ireland* (Oxford, 1997), 71–2.

[66] *Ibid.*

the focus of the war in Catholic consciousness in Ulster. The public display of memory in the social space of towns and cities that mimicked the geography of the Orange Order's parades would have serious consequences for the creation of a collective memory that would transcend the sectarian divisions of this part of the island. The introduction of mini-parades by the Orange Order on 1 July, to commemorate the Somme, would institutionalise the interpretation of the war in Ulster and codify it in ways which deviated from the official day of national remembrance on 11 November. The idiom of the memory makers' calendar in Ulster underscored Protestant desire to celebrate British identity while simultaneously marking this identity through localising discourse. The dispute over the routing of the parade from Drumcree church to the centre of Portadown in recent years is, ironically, a conflict over the use of public space for the remembrance of the Battle of the Somme rather than the Battle of the Boyne. The fact that this has no significance for either side in the dispute reinforces the aggregation of the two events into a single motif of identity.

Belfast's 1919 parade

We are today in Belfast joining with our fellow citizens of the British Empire in expressing our heartfelt thankfulness to Almighty God for complete triumph over an arrogant and remorseless enemy.[67]

The choice by Belfast to postpone its Peace Day celebrations until August 1919, so as not to conflict with the Orange parades of July, provides an opportunity to examine how the war was modulated in Ulster a month after the national celebrations. While 9 August had been designated for the civic celebrations by Belfast's (unionist) city council in July, decisions to participate in the parade were largely reserved until August. Two days before the peace parade, at a special meeting in Derry of the local branch of the Nationalist Veterans' Association, led by Alderman J. M. Monagle, the attitude of nationalist ex-servicemen towards the parade was debated. Based on the advertising literature for the parade, which focused on the role of the Ulster division, a motion proposed by two ex-soldiers was unanimously passed. It stated: 'We believe it is being held for political motives, which are contrary to the rights and principles of freedom we, Nationalists, fought for, and we call on all Nationalist ex-servicemen to refrain from participation in it.'[68] For Ulster men the motives underlying commemoration conjured up variable responses. The editorial of the *Belfast Newsletter* vehemently denied political motives, commenting that such allegations emanated from 'those whose acquaintance with loyalty is not of the most intimate nature'.[69] In a side-swipe at the capital city's efforts the previous month, the paper claimed that 'No such demonstration as that which will salute His Majesty's representative [Lord French] has ever been organised in Ireland, nor, we imagine, have so many people ever been brought together for

---

[67] Editorial, *Belfast Newsletter*, 9 August 1919.   [68] Cited in *Freeman's Journal*, 9 August 1919.
[69] Editorial, *Belfast Newsletter*, 9 August 1919.

entertainment at one time.'[70] Those eligible to apply to participate in the parade included the following: Ulstermen who served during the war; all others who served in Ulster units; all demobilised officers and men now residing in Ulster and all Ulster ladies who took up service during the war and all others with service now living in Ulster. While the organisers anticipated about 20,000 applications, in the final account numbers swelled to somewhere between 30,000 and 36,000 participants.

The parade was organised around a route of approximately 4 miles beginning in north Belfast and ending at Ormeau Park, south east of the city centre (Figure 19). The symbolic nexus of the route was the City Hall where Ireland's viceroy, Lord French, took the salute. He had made the journey by car from Dublin to attend the event. City Hall, located in Belfast's central square and housing the city's Council, formed the local centrepiece of the parade. Behind the saluting point, in stands specially erected for the occasion, were invited guests. These included Field Marshal Sir Henry Wilson and other senior military personnel, Unionist Members of Parliament such as Sir Edward Carson (MP for Belfast Duncairn), Mr G. Hanna (MP for East Antrim), city councillors headed by the Lord Mayor Mr J. C. White and a variety of local members of the aristocracy.[71] Sir Edward Carson had arrived in Belfast the previous day to unveil a roll of honour at the Workman, Clark and Co. shipyard, where some 2,600 men had volunteered for service. In addition to that official duty, Carson attended an important meeting of the Standing Committee of the Ulster Unionist Council where he delivered a speech on the current political situation. At the meeting it was resolved to convene a meeting of the Unionist Councils and to revive various political organisations including the Unionist Clubs that had been dissolved during the war. It was also agreed to commemorate Covenant Day with religious services throughout Ulster on 28 September. Carson addressed a series of political demonstrations during this time[72] and the convergence of regional commemoration with national political objectives was striking. The death of soldiers along the battlefront was not ideologically separated from the status of the union. In a tribute to the men who had fallen during the war, Carson declared:

I myself came in at the head of the first Volunteer regiment which formed the nucleus of the Ulster division and marched with them to the recruiting office ... I never doubted that they would acquit themselves on the field of battle as great soldiers, loyal to their king and country, and that our old motto of the province, 'No surrender', would be the guiding ideal when they came in contact with the Hun aggressor.[73]

Clearly identifying a lineage with the will to defend the nation, Carson's comments trace an ideological link between pre-war conditions in Ulster and the conclusion of the war. Repetition, as a keystone in the fashioning of memory, is central in his statement.

---

[70] *Ibid.*   [71] *Irish Independent*, 11 August 1919.
[72] *Irish Times*, 9 August 1919.   [73] *Belfast Newsletter*, 11 August 1919.

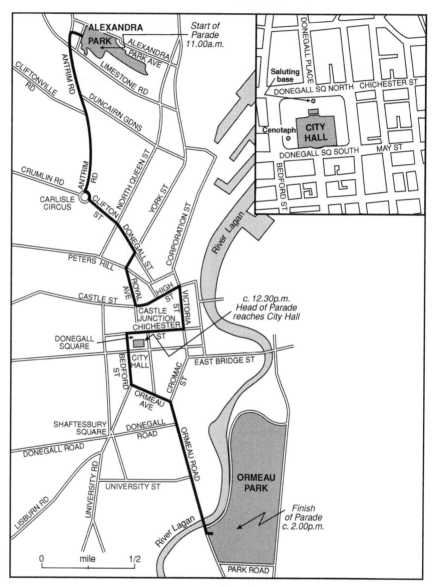

Figure 19 Route of Belfast's parade August 1919

The parade began at 11.10a.m. departing from Alexandra Park, north of the city centre, and travelling along the Antrim Road and into the city centre. The Royal Navy and auxiliary forces were given the post of honour. While most of the demobilised men were from Ulster, many from the Munster Fusiliers and Connaught Rangers were also present. Flags decorated the main thoroughfares. They included

Figure 20 Belfast Cenotaph

a Japanese flag and an American flag. The Irish Rifles (11th battalion) was led by Captain C. C. Craig, MP for South Antrim, and he was loudly cheered when recognised. Along the Antrim Road, 'Tipperary' was sung by the crowd and played by the band of the Irish Guards. Wounded soldiers were conveyed in motor vehicles, charabancs and lorries, prompting the *Belfast Newsletter* to comment that 'they were so merry and boisterous, and so resolutely bent on sustaining the festive spirit of the occasion, that commiseration would have been glaringly out of place'.[74]

While spectators lined the entire route, the real concentration was at the viceroy's saluting point in front of the Queen Victoria statue at City Hall. On the west side of City Hall there was placed a cenotaph to the memory of the dead (Figure 20).

[74] *Belfast Newsletter*, 9 August 1919.

A guard of honour composed of four soldiers, with bowed heads, stood around the cenotaph. Many wreaths were laid at the cenotaph and the whole area was ablaze with bunting. Behind lines of armed soldiers who managed the route of the parade was a thick fringe of spectators. At 12 noon the parade reached the saluting point and it took 3 hours for it to pass. When the 36th Ulster Division reached the saluting point there was a 'prolonged outburst of cheering. The men made a very fine show. The greater proportion of the men was in civilian attire and they marched with true soldierly bearing. Many wore the ribbons belonging to certain decorations and the spectators' hearts filled with pride as they gazed upon those men who have brought so much credit to their Province and to Ireland.'[75] After the procession had passed, the Guards Band played the National Anthem and the lord lieutenant was cheered. The greater public enthusiasm, however, was extended to Sir Edward Carson, where the crowds sought handshakes and a speech that he willingly supplied. He said, 'I never was prouder of Ulster and her heroes than I was today. May God bless Ulster and may God bless the King.'[76] Afterwards there was a luncheon at City Hall for Lord French and the other invited guests, while at Ormeau Park marquees had been erected to feed and entertain the thousands of soldiers who took part in the parade. During the luncheon at City Hall the viceroy made a speech of thanks to the people of Belfast, noting that:

In spite of the absence of military weapons and uniforms the spectacle was magnificent with a grandeur which no military environment could have created. It was rendered so by the bearing of victorious soldiers who had fought their way through bloodstained paths by deeds of unparalleled heroism over the bodies of the flower of their country's manhood to victory.[77]

Focusing on the fact that the soldiers were volunteers, Lord French commented on the free will they exercised in joining the armed forces. The manifestation of patriotic duty was calibrated rather differently in Ulster than in other parts of the Ireland. The lord mayor went on to honour His Majesty's representative in Ireland 'who after a rite of toil and danger, was now in stormy and difficult times, the staunch guardian of peace and order among them'.[78] The fragile political position of Ireland in 1919 was ironically reminiscent of the summer of 1914. While the course of the war may have blurred political allegiances in Ireland, divisions were to resurface quickly once the conflict was concluded.

The spectacle in Belfast, although sharing some of the characteristics of the parades held elsewhere in the country, differed in important respects. If, as Barthes suggests, the function of spectacle is 'the exaggeratedly visible explanation of Necessity',[79] the scale of the event in Belfast was indeed exaggerated. The length of the route, number of participants and spectators exceeded what happened elsewhere in the country. One commentator stated that, 'As a spectacle the march was magnificent, but its true significance lay in the powerful appeal which it made

[75] *Irish Times*, 11 August 1919.   [76] *Ibid.*   [77] *Ibid.*
[78] *Ibid.*   [79] Barthes, 'The world of wrestling', 20.

to the emotions.'[80] These emotions were aroused through the iconography of the parade and through the narrative that subsequently came to reflect it. The links between physical prowess, 'race' and regional identity were repeatedly used to represent the parade in Belfast. The *Belfast Newsletter* employed phrases such as 'brave clansmen of Ulster', 'our sturdy population' and a 'great and imperially-minded race' to situate Ulster's position in the larger theatre of war.[81] The *Irish Times* also remarked that the celebrations in Belfast were 'a striking demonstra-tion... of Ulster's loyalty and adherence to the Throne and the Constitution. It was at the same time convincing proof of the noble part played by the Northern province in the Great War.'[82] The nomenclature of individual and 'national' hero-ism reinforced in the minds of Ulster Protestants, in particular, the significance of their sacrifice in the broader political arena of Anglo-Irish relationships. The more nationalist-leaning newspaper, the *Irish News and Belfast Morning Post*, offered more reserved comment on the day's events. It stressed that the lack of splendour and pageantry associated with such commemorations was due to 'the proportion of those in the ranks who wore ordinary civilian dress', and the disorganisation at Ormeau Park, where many soldiers who had travelled long distances received no food or drink.[83] Similarly, although making front-page news in the Dublin-based *Saturday Herald*, the column was a brief and descriptive account of the day's events. Responses to parades varied greatly geographically and politically and the exceptionalism, which characterised Ulster's interpretation of its role in the war, continued a discourse which focused on Ulster's difference to the remainder of the island. In other Ulster towns which held parades, such as Antrim, Bangor and Lisburn, the holiday passed off without incident. In Enniskillen, despite the refusal of the nationalist Urban District Council to take part in the celebrations, the holiday was observed by both Catholics and Protestants.[84]

Conclusion

Irish men and Irish women engaged in large numbers in the Great War and endured comparable hardship to that of other national groups immersed in the conflict. The consummation of Allied victory expressed through national Peace Day celebrations and annual Remembrance Day spectacles amplified Ireland's equivocal response to that effort. Tom Kettle, a Member of Parliament, an academic and a supporter of Home Rule, realised the ambiguity of his position as a soldier on the Western Front. Kettle was acutely aware of how the public memory might subsequently be fashioned: 'These men [of Easter 1916] will go down in history as heroes and martyrs, and I will go down – if I go down at all – as a bloody British officer.'[85] Although a commemorative tradition had been established in the previous couple

---

[80] *Belfast Newsletter*, 11 August 1919.   [81] *Ibid.*
[82] *Irish Times*, 11 August 1919.   [83] *Irish News and Belfast Morning Post*, 11 August 1919.
[84] *Belfast Newsletter*, 21 July 1919.   [85] Cited Boyce, 'Ireland and the First World War', 51.

of centuries from mass funerals, Orange Order parades or centenary celebrations of the rebellion of 1798, in the years immediately after the Armistice the question of how Irish people would make sense of their role in the First World War was deeply contested. The evidence suggests that it was not simply a desire to expunge the memory of war completely from public consciousness by nationalists and to hyperbolise it by unionists that governed the form of remembrance in Ireland. In some important respects, the difficulty lay in the fact that for many Irish people the war of ideas and ideological conflict was not over. The on-going dispute over Home Rule that plagued pre-war attitudes towards the war quickly resurfaced in its aftermath. The drama of 1916 in both the Somme and Dublin affected all shades of political opinion and informed in contradictory ways the staging of a public spectacle of remembrance.

In Barthes' discussion of spectacle its potency resides precisely in 'the popular and age-old image of the perfect intelligibility of reality... in which signs at last correspond to causes'.[86] The use of spectacle in Europe to construct a post-war memory, even among the victorious, could not rely on such certainties. The meaning of the war could not be staged so easily, perhaps because the actions of the war itself were beyond the conventional parameters of intelligibility. Paul Fussell's disarming contention that, 'In the Great War eight million people were destroyed because two persons, the Archduke Francis Ferdinand and his Consort, had been shot',[87] exposes the paradox of cause and effect. Nevertheless combatant states expended enormous physical and financial energy in trying to make sense of just that, by freeze-framing the war in the public consciousness through peace day celebrations and subsequently through annual remembrance Sundays.

In the case of Ireland, this approach to remembrance presented deep-seated contradictions for participants and public alike. In the capital city and in many Ulster towns a spectacle could be staged with relative success (at least in daylight hours), while in other places, more remote from the administrative centre of the island, support was more muted. The representation of four consecutive years of war through a single street event underlines the contention that the war was popularly mediated more through spatial than temporal categories. Military units configured into a public spectacle, marching along streets lined by spectators, disguised the fact that the war was a sequence of conflicts fought on different sites, at different times, across Europe. Veteran soldiers became the undifferentiated representatives of a moral order (the just cause), and their willingness to serve (notwithstanding conscription) deserved the symbolic thanks of the state and the public at large. In Ulster, this expression of thanks was particularly compressed around the Battle of the Somme and the 36th Ulster Division.

Public memory, more generally, was cultivated through the spaces in which the parades took place and the formal iconography (flags, uniforms, anthems) surrounding them. In the Irish case I have suggested that the population did

---

[86] Barthes, 'The world of wrestling', 29.   [87] Fussell, *The Great War and modern memory*, 8.

discriminate between the veteran and the regular soldier; and that the populace differentially read the icons of legitimisation. The material representation of the war through flags, uniforms, artillery and the discursive representation of loss through speeches and salutations were differentially received among the populace. Attempts to have the parades read uniformly by the public were difficult in post-war Ireland because the very symbolism employed had wider meanings in the context of 1919 and was contradictory in its effects. Like Barthes' world of wrestling, the parades may have sought to offer an intelligible basis to suffering but the commemoration of one war in the shadow of another set in stark relief the ambiguity between the past and our reading of it. Local circumstances could not be submerged totally under national narratives and the parading of peace in an un-peaceful environment brought into focus the resilience of domestic preoccupations even within a universalising rhetoric of remembrance. While the parades of 1919 provided the initial step towards rendering a calendar of memory and offer us a snapshot of Irish responses to the war, the creation of more permanent landscapes of memory indicate how these mixed sets of responses to the war were carried through in the decade immediately following the Armistice. In Chapter 4 the situating of memory around specific sites of mourning and the debates underpinning such public geographies will be the focus of our attention. If parades formed a transient albeit spectacular expression of public remembrance, memorial sites formed an embedded and permanent reminder of the pain of war.

4

Sculpting memory: space, memorials and rituals of remembrance

Chapter 3 addressed the spectacle of remembrance at the Peace Day parades, yet the material basis of much commemorative energy focused on the construction of permanent spaces and artefacts of memory on both the war and Home Fronts. The mapping of the casualties of war through the making of memorial sites aroused widespread debate for the victorious and the defeated. It formed part of a larger on-going process in the redesign of public space through monumental architecture and statuary. War memorials thus exist in tandem with a suite of other markers, which map the cultural and historical identity of cities, regions and nation-states. Wagner-Pacifini and Schwartz suggest that 'Memorial devices are not self-created; they are conceived and built by those who wish to bring to consciousness the events and people that others are inclined to forget.'[1] In particular, in post-war Ireland, the politics of remembrance and amnesia were interrelated in complex ways. With the establishment of the Irish Free State and the partition of the island, the use of public space to articulate a version of the past and a vision of the future was a highly disputed issue on both sides of the border. The claim that 'Statues or monuments to the dead...owe their meaning to their intrinsic existence...[and] one could justify relocating them without altering their meaning',[2] is to ignore the historical moment and geographic specificity of debates undergirding the building of war memorials and to overemphasise their transcendental qualities. While Maya Lin's quasi-abstract Vietnam Veterans' Memorial in Washington DC is a powerful icon of remembrance, its meaning is at least partly defined by its location on the Mall in America's capital city and its links to US foreign policy and geopolitical discourses. The interpretation attending the monument might be rather differently viewed if its home was in Ho Chi Minh City.

In the context of First World War memorials, different national states adopted different approaches to commemoration and the debates surrounding the construction

---

[1] R. Wagner-Pacifini and B. Schwartz, 'The Vietnam Veterans' memorial: commemorating a difficult past', *American Journal of Sociology*, 97 (1991), 382.

[2] Nora, 'Between memory and history', 22.

of sites of memory varied. In terms of official memorials, Hynes has suggested that they represent 'acts of closure... that bring war to a grand and affirming conclusion'.[3] He contrasts memorials with literature, paintings and other cultural productions which were more contingent in their representation of war and their commemorative support of it. Kertzer has reminded us that the rituals surrounding memorialisation 'can serve political organisations by producing bonds of solidarity without requiring uniformity of belief'.[4] In France, for example, the issue of commemoration converged around two areas of dispute. One related to the use of religious or secular iconography in the geography of monument design, the other focused on 'the negotiation of local and national claims to memory of the dead'.[5] The French government agreed, where possible, to pay for the return home of soldiers' bodies and frequently local memorials erected in towns and villages named individual soldiers killed in the community. In the United States, despite the War Department's desire to bury American soldiers in the war cemeteries of Europe and inscribe a permanent reminder of the United States' role in the world political order on that landscape, American women refused to have their loved ones treated solely as servants of the state. Eventually more than 70 per cent of the bodies of soldiers were repatriated after a public discussion of individual rights of citizenship, the state's rights, gender and the appropriation of commemorative practice.[6] In Australia, memorials to the war are dedicated to all who served in the war, not just the war dead. The dead and the living are merged in the Australian context and this partly relates to the controversy surrounding conscription in Australia.[7] It also is an acknowledgement of the deadening effect of war even on the lives of living veterans. It places the corporeal body, whether living or dead, at the centre of the commemorative narrative and shifts the focus of the discourse of war away from the loss of individual combatants to the broader cost of war to a society in general.

In respect of Britain's response to its war dead, a heated debate took place, anchored around the twin issues of where and how these soldiers could be best officially commemorated. Before the Great War, the construction of official memorials and burial sites to the rank and file war dead was comparatively rare. But the sheer volume of casualties in the Great War and the fact that the army comprised largely volunteer recruits rather than professional soldiers altered entirely the terms of reference. The geographical scale of fighting, the number of soldiers killed, and the manner in which they were killed, quickly necessitated action on

---

[3] Hynes, *A war imagined*, 270.
[4] D. Kertzer, *Ritual, politics and power* (London, 1988), 67.
[5] D. J. Sherman, 'Art, commerce and the production of memory in France after World War I' in Gillis, *Commemorations*, 188.
[6] G. K. Piehler, 'The war dead and the gold star: American commemoration of the First World War' in Gillis, *Commemorations*, 168–85.
[7] K. S. Inglis and J. Phillips, 'War memorials in Australia and New Zealand: A Comparative Survey', *Australian Historical Studies*, 24 (1991), 171–91.

the registration and burial of the dead. Under Fabian Ware, the Graves Registration Commission began this task and by May 1916 50,000 graves had been registered and 200 battlefield cemeteries were under construction in France and Belgium. Whilst initially it had been thought that corpses would be repatriated to Britain once the war ended, the horrendous volume of casualties the war had generated brought to light the immense logistical and moral dilemmas that would be raised by such a strategy. Consequently, before the war had ended debates began to take place on how to remember and represent the dead abroad.[8] Once it was decided to bury the deceased along the Western Front, 'the state poured enormous human, financial, administrative, artistic and diplomatic resources in to preserving and remembering the names of individual common soldiers'.[9]

In a complex exchange of views, the Imperial War Graves Commission favoured the construction of simple uniform headstones (instead of crosses which were seen as too Catholic a symbol to embrace all the religious and non-religious participants in the war). Opponents to this proposal objected to the uniformity, impersonality and secularism implied by the design.[10] After heated parliamentary debates and vitriolic exchanges in the letters pages of the national press, the central issue to emerge concerned the ownership of the bodies and the legitimacy of the state to manage their burial and remembrance. By 1920 this issue was resolved: 'The war dead were henceforth public property, and their commemoration was to be organised not by individuals in private burial places but by an official bureaucracy.'[11] An official landscape of remembrance was to be inaugurated along the Western Front and to this end thousands of Portland stone headstones were transported across the English Channel and erected in a series of specially commissioned war cemeteries in Belgium and France. By 1930, there were over 540,000 headstones erected in 891 cemeteries which were designed to mirror many of the quintessential elements of English-style landscape architecture. Although buried abroad, there was an attempt to domesticate the landscape in ways in keeping with the contemporary tastes of those at home.[12] In contrast to the United States' desire to map their geopolitical interests in Europe, for Britain the interring of the dead on foreign soil provoked a desire to transfer a little bit of Britain overseas and to translate the landscape of home to foreign soil. As Halbwachs has reminded us, 'every collective memory unfolds within a spatial framework. Space is a reality that endures: since our impressions rush by, one after another, and leave nothing behind in the mind, we can understand how we recapture the past only by understanding how it is, in effect, preserved by our physical surroundings.'[13] The physical surroundings

[8] Heffernan, 'For ever England', 293–324; Gregory, *The silence of memory.*

[9] T. W. Lacqueur, 'Memory and naming in the Great War' in Gillis, *Commemorations*, 155.

[10] B. Bushaway, 'Name upon name: the Great War and remembrance' in Porter ed., *Myths of the English*, 136–67.

[11] Heffernan, 'For ever England', 305.   [12] Morris, 'Gardens "For ever England"', 410–34.

[13] Halbwachs, *On Collective memory*, 140.

of the battlefields became the ideal sites for the representation and preservation of the war dead.

The symbolic keystone of remembrance at home was the building of the catafalque in Whitehall and the burial of the unknown soldier in Westminster Abbey: 'the unknown warrior becomes in his universality the cipher that can mean anything, the bones that represent any or all bones equally well or badly'.[14] Not all interests, however, were satisfied with the Cenotaph, and the *Catholic Herald* attacked the monument as 'nothing more or less than a pagan memorial [which was] a disgrace in a so called Christian land'.[15] In an attempt to take the theological wind out of the sails of the Anglican Church, the Catholic Church sought to reinforce their position as the true homeland of Christian morality, tradition and iconography. In towns and villages across the United Kingdom smaller-scale memorial spaces matched those in the capital. If the memorial sites at home represented the abstract and communal articulation of loss, the war cemeteries along the Western Front represented the physical and more individualised evidence of the large-scale destruction of soldiers in the First World War. As Heffernan has reminded us: 'The official commemoration of the war dead is articulated around a complex geography, combining domestic ceremonials from which the dead are excluded and a vast network of overseas memorials and cemeteries where individual soldiers are recalled and their actual remains interred.'[16] Pilgrimages to these sites remain as popular today as they did for the families of the dead in the years immediately following the conflict.

In Ireland, the debates surrounding the memorialisation of the dead shared some of the same concerns as those in other states while at the same time localising the issues central to these discussions. This chapter will examine, in some detail, the debate in Ireland about the choice and use of public space to establish a national war memorial. The chapter will also consider the role of memorials in a regional context. The significance of memorials to Ulster Protestants both at home and on the Western Front will be examined in light of attitudes expressed elsewhere around the island. Finally, I will suggest that the sculptural mapping of the war in the Irish Free State was informed by a contested reading of the literal and symbolic place of the war in Irish historiography. This contrasted with the emphasis placed on the war within the existing commemorative lexicon of Ulster's representation of its past. As Lefebvre noted, the meaning spaces provoke is the product of dispute as different 'groups or classes [seek] to appropriate the space in question'.[17] Not only then were spaces of memory also sites of mourning, innocent reminders of the pain of war, they were important symbolic reference points for the cartographies around which the historical record on the island would be chalked.

---

[14] Lacqueur, 'Memory and naming', 158.   [15] Quoted in Gregory, *The silence of memory*, 199.
[16] Heffernan, 'Forever England', 295.
[17] H. Lefebvre, *The production of space* (Oxford, 1991), 57.

Irish National War Memorial

The confluence of Peace Day celebrations, proposals to create permanent memorials to the dead and the establishment of a central branch of Comrades of the Great War in Dublin in the summer of 1919 all attested to the immediate post-war attempt to provide a solid footing for commemoration in Ireland. Lord French opened the central branch of the Comrades of the Great War at 42 York Street, Dublin just prior to the 19 July public parade. Stressing that the organisation's membership was made up of 2,500 men who had served in all sections of the service during the war, Lord French emphasised that 'politics are eschewed within its walls', and further noted that it was 'the only body of its kind which is absolutely untouched with any political colour, and it is for that reason that I am personally able to give it my wholehearted support'.[18] A memorial to the Irish killed in the war was mooted as early as late 1918. But it was not until 17 July 1919 that a general meeting was convened at the Vice-Regal Lodge under the presidency of Lord French and that the initial proposals got under way. Although Lord French would have liked to have had a committee established earlier, he commented that: 'Nothing, however, could be done until the whole of loyal Ireland was brought into council', and although a peace celebration would be held in Dublin on 19 July, he noted: 'It is right that our thankfulness should take a permanent form, so that those who come after us may remember the struggle which their predecessors had to keep intact the glorious principles upon which our Imperial life is based.'[19]

The initial tentative proposals for the permanent memorial comprised a Soldiers' Central Home – Great War Memorial Home – which would provide accommodation and entertainment space for soldiers and ex-servicemen passing through the city of Dublin. *Ireland's War Memorial Records* would be published, listing all of Ireland's war dead and they would be housed at the site. The estimated cost of the entire scheme was £50,000. The proposal was subject to debate at this meeting and the lord chancellor of Ireland offered his support for the project. The provost of Trinity College, Dr Bernard, insisted that the memorial should be both permanent and one that all Ireland could support. Although favouring a location in Dublin the provost was keen that the memorial be a 'symbol of unity' on the island. Captain Stephen Gwynn similarly supported the idea of a memorial but was anxious that the purpose of any proposed memorial be made explicit to the public and that soldiers' organisations be consulted on the style and idea behind any memorial. Representatives from Ulster, although supportive of the principle of a memorial gesture, made the committee aware that they had already begun some commemorative work. This meeting established the basis for planning a memorial in Dublin and a committee of over one hundred people from varying walks of life was selected and offices were acquired at 52 Dawson Street to plan and regulate fund-raising activities.[20]

[18] *Irish Times*, 19 July 1919.   [19] *Irish Times*, 18 July 1919.   [20] *Ibid.*

Although many subscribers to the fund supported the proposal to erect a soldiers' home in Dublin, the military authorities came to a decision that a home would be unsuitable. They did not approve of serving soldiers mixing in a club with ex-servicemen who were no longer subject to military discipline and control. In addition, the political climate in Ireland in late 1919 altered the committee's thinking and the memorial home was abandoned. After the signing of the Treaty in 1921 it became inconceivable to establish a home for British soldiers on Irish soil. The committee continued in its efforts to raise funds notifying the public that:

The memorial is to stand in the capital of Ireland, and is destined to keep alive in the hearts of the Irish people for ever the glorious memory of their heroic dead, who in the world's greatest struggle for freedom died for the honour of Ireland.The memorial will be representative of every class and creed. To make the memorial worthy of its lofty object, and a really national one, it is essential that contributions be received from all parts of Ireland.[21]

Alternative proposals began to be aired: the construction of a cenotaph, arch, gate or fountain in some prominent part of central Dublin; a memorial hall containing the records of dead soldiers; a park, model villages and workshops for disabled ex-soldiers, halls or clubs for ex-servicemen and charitable funds to be dispensed to war veterans. In addition, the committee decided to contribute funds for the erection of stone crosses to replace the wooden crosses in memory of the Irish Divisions who served in the battlefields of Flanders, France and Gallipoli.

From 1919 to 1923, most of the committee's energies were devoted to fund-raising and compiling the necessary information to produce the bound volumes containing the names and personal details of every Irish soldier killed in the war. Copies of the volumes were presented to the king, the pope and to the Protestant St Patrick's cathedral in Dublin. As the Irish Civil War (1922–23) came to a close towards the end of 1923, the War Memorial Committee again began to reconsider proposals for a commemorative memorial. The workings of the Committee should not be separated from the wider political changes taking place. The British Legion in Ireland informed the committee that it had 'the universal wish of the Council that the memorial should take the form of a statue, obelisk or cenotaph of exceptional beauty and grandeur, sited in some central part of the City of Dublin'.[22]

Now that the memorial home idea had been abandoned, more specific proposals for a public monument were articulated. The Committee suggested that it acquire the private park in Merrion Square in Dublin and that it convert the square into a public garden named the Memorial Park (Figure 18). The park would contain a suitable monument in its centre and at each corner of the park would stand an emblematic entrance gate. The park would also be laid out in flower and shrub beds, which, when completed, would be handed over to the state for maintenance. Funds in the region of £46,000 had been raised to support the project. Given the

---

[21] Quoted by Senator Jameson, a trustee of the Irish National War Memorial, Seanad Éireann, *Official Report*, viii, cols. 422–3, 9 March 1927.

[22] *Ibid.*, cols. 424–5.

legal complexities of acquiring the land from private ownership, the proposal would require the passing of a Private Bill in the *Oireachtas* (legislature). On 28 March, 1924 a meeting of the General Committee was convened and it approved this proposal. In consultation with the Merrion Square Commissioners, Lord Pembroke, Dublin Corporation and the High Court, the Merrion Square (Dublin) Bill 1927, a Private Bill was presented for second reading in the Senate in March 1927. It had taken over seven years for a firm proposal to commemorate Irish soldiers of the Great War to reach the Irish parliament. Internal disagreements and the external political context both played a role in delaying action on the proposed memorial. Even when it reached the Irish Senate, the debate revealed that the merits of the project continued to be debated both from within the ranks of the War Memorial Committee, the residents of Merrion Square and the wider society.

Although much of the debate in the Senate related to matters of parliamentary procedure concerning the reading of a Private Bill, matters of principle also arose in relation to the project. Senator Sir Bryan Mahon (commander of the 10th Irish division during the war) raised objections to the proposal for Merrion Square. While he had no objection to the construction of an individual memorial, he noted that 'When honouring the dead we ought not to forget the living . . . I would suggest erecting a suitable memorial, and that the balance of the money, if any, be devoted to the benefit of ex-servicemen.'[23] The expenditure of over £40,000 on a single site of remembrance seemed excessive to Senator Mahon and he further expressed concern about its situation. The location in central Dublin, in his view, would cause problems during the 11 November commemorations. Merrion Square would not have the capacity to host the large assembly that gathers on Armistice Day without causing major traffic disruption and inconvenience in the city, as had occurred in previous years when the commemorations were held in College Green and St Stephen's Green in the city centre. In addition, the 11 November ceremonies were also threatened by disturbance from those opposed to Armistice Day. Senator Mahon conjectured, 'Is the heart of Dublin, under the very walls of the seat of Government, the place in which to take that risk and the risk exists, as everyone in this House knows?'[24] Even proponents of the memorial could not ignore the geographical proximity of Merrion Square to the parliament building (Leinster House: see Figure 18).[25] Senator Mahon suggested that a memorial be located in the Phoenix Park,[26] a large public park where previous Armistice Day celebrations

---

[23] Sir Bryan Mahon, *ibid.*, cols. 413–14.   [24] *Ibid.*, cols. 414–15.

[25] Leinster House was built by the architect Richard Cassels in 1745 for Lord Kildare, duke of Leinster. It was the largest private residence in the city. The Royal Dublin Society (RDS) purchased the house in 1815 and in 1922 the Irish Free State bought the house from the RDS for £68,000 to accommodate the Irish parliament (Dáil and Seanad, upper and lower houses). While the front of the house faced Kildare Street, the back of the house faced Merrion Square.

[26] In the Phoenix Park, just west of the city centre along the banks of the River Liffey, was located the home of the viceroy. The original house, built in the eighteenth century, was extended and renovated in the nineteenth and was frequently used to welcome various monarchs, including George IV,

had taken place, and which would avoid causing disruption to the commercial life of the city or causing public embarrassment in the event of 'trouble'. The fact that the Wellington monument was also located in the Phoenix Park may also have added a dimension of continuity to Irish military history in the imperial realm.

Senator Sir William Hickie, ex-serviceman, member of the British Legion in Ireland and member of the Council of the National War Memorial, also raised reservations about the Merrion Square proposal. While the legion had initially supported the proposal, Senator Hickie noted that, on reflection, legion members held the view that the square could not accommodate the 50,000 or more people who attend Remembrance Day ceremonies without damaging the park and disrupting city life. The fact that a portion of the park was to be reserved as private tennis courts for residents of the square did not win the approval of ex-servicemen. Ironically then, Senator Mahon, a signatory of the Bill, in the debate publicly opposed the Bill.[27]

Other senators also expressed concern about the possibility of political disruption at Remembrance Day ceremonies especially when the spatial proximity between the memorial and the houses of parliament was considered. On this issue Colonel Moore, a government representative at previous ceremonies, noted:

Under present circumstances it [Merrion Square] would be almost certain to raise adverse discussion, and perhaps even serious trouble . . . I am sure that those who had relatives killed in the war would not wish that there should be scrimmaging and trouble . . . I, at all events, would feel very sore and bitter if any such thing happened over relatives of mine.[28]

Such debates about the use of public space for personal mourning were not confined to Ireland. Sherman has noted in the context of France that, 'In a cemetery, the bereaved could mourn in peace; on a public square a monument stood more emphatically for the community's will to commemorate.'[29] Senator Oliver St John Gogarty, writer and surgeon, claimed that only about 10 per cent of the citizens of the Free State were especially interested in a memorial and that

The centre of the city is not . . . the best platform for annual panegyrics. A war memorial is a comfortless thing. I do not know of any greater monstrosity than the Wellington monument. There is not shelter on it for a sparrow. If the money subscribed is to be turned into stone, the best thing would be to provide houses for ex-servicemen.[30]

The tension between commemorating the dead and providing for the living animated many debates about post-war commemoration in Europe. War memorials, cast in stone, by no means aroused universal support. Their potential to become

Queen Victoria, Edward VII and George V. After independence the house was handed over to the new government and was used for some time as the residence of the governors general, before becoming the official residence of the Irish president in 1938.
27 Sir William Hickie, Seanad Éireann, *Official Report*, viii, cols. 417–419.
28 Colonel Moore, *ibid.*, col. 419. 29 Sherman, *The construction of memory*, 218.
30 Dr Gogarty, Seanad Éireann, *Official Report*, viii, col. 420.

innocuous sites, reclaimed annually for public ritual to commemorate the dead, competed with demands for the living. Notwithstanding the political context of post-war Ireland, which provides additional nuance to the debate, utilitarian versus symbolic gestures to collective memory conjured diverse responses on how a debt of gratitude would be most sympathetically delivered. That houses for veterans would not provide a public site for the ritualisation of commemoration exposes the larger tensions between attempts at the promotion of historical memory in the public sphere and the accommodation of private grief and reconciliation.

Senator Jameson, in his address to the house, sought to reflect the wishes of the subscribers to the memorial fund. While he acknowledged that the site would be unsuitable for annual Remembrance Day parades, he insisted that this was not the primary purpose of the proposal. Indeed, he suggested that a cenotaph could be erected in the Phoenix Park, at a small cost, for that purpose. The park, he suggested, served a larger public interest and the committee would not desire that such a large sum of money be spent on an individual monument. The Merrion Square site would become a public utility serving the population of the inner city as a leisure space and in so doing removing it from private ownership for the sole pleasure of the residents of the square. It would serve a civic as well as a commemorative role. On these grounds, Senator Jameson supported a second reading of the Bill.[31]

On behalf of the residents of the square, Senator Barneville claimed that they were opposed to the Bill, particularly as a site for ritual remembrance:

There is no doubt there are certain points of view in this country which many of us regard as already old-fashioned prejudice; as bigoted, but still there is no denying that these points of view are there, and held very strongly indeed. We feel ... that these points of view, particularly at times of political disturbance, would lead to commotion, demonstration and counter-demonstration in our Square ... such disturbances ... would lead to a depreciation in the value of our property.[32]

The possibility that the site would become a focus for ideological conflict prompted the negative response from some of the local residents. By contrast, Senator Yeats claimed that he would like to see the square developed as a public park for Dublin's children. As a memorial site, however, Yeats surmised that its function would be short-lived: 'I do not think we should take too seriously the interests, the fancies or desires of even those admirable men who want a great demonstration upon Armistice Day. Armistice Day will recede.'[33] While supporting the idea of a monument in the square listing the fallen soldiers for the benefit of their descendants, Yeats queried the underlying purpose of public commemoration. In his view, public demonstrations are orchestrated and do not flow naturally from ex-servicemen and their relatives despite claims made by senators of military background. If the state or the leaders of the ex-servicemens' organisations did

[31] Mr Jameson, *ibid.*, cols. 431–2.   [32] Dr Barneville, *ibid.*, col. 443.
[33] Dr W. B. Yeats, *ibid.*, col. 444.

not organise commemorations in the square, Yeats claimed, there would be no political conflict and the square would be of benefit to the health and wellbeing of the city's citizens for generations. He added, however, 'I do not believe that in 100 years any monument erected now will be very important.'[34] For Yeats the central issue was the provision of a public recreation area for Dublin's adults and children. His commitment to the role of Irish soldiers in the Great War was far weaker. The distinction drawn between the provision of a memorial for individual contemplation contrasts with his antipathy to public ritual. The Bill, however, narrowly passed its second reading with the chairman casting the deciding vote.

Ironically the supporters and opponents of the scheme proposed in the Bill during the Senate debate contrast with what might be expected. Representatives of ex-servicemen's associations opposed the proposal, while some with more nationalist political leanings favoured it. This in part revolved around the central issue of the precise purpose of the park. Was it to serve the civic needs of the city as a public park? Was it to be a solemn site of commemoration with annual rehearsals of Remembrance Day rituals? Was it to serve individual or collective needs?

If the Senate debate exposed some of the opposing attitudes towards commemoration of the Great War, the Dáil debate indicated more clearly some of the government's objections to the site of Merrion Square. The vice-president of the Executive Council, Deputy O'Higgins, made the opening statement to the Bill. Both from a personal perspective and as the conveyor of the views of the Executive Council, O'Higgins voiced strong opposition to the proposal on two grounds. O'Higgins first claimed that the subscribers to the fund were opposed to the Bill and that this was reflected in a letter written to the *Irish Times* by Lord Glenavy, a member of the original committee established by Lord French at the meeting in the Vice-Regal Lodge in 1919. In his letter, Lord Glenavy suggested that once the memorial home proposal had to be abandoned, the proposal to convert Merrion Square into a public park – a municipal project to serve the citizens of the city – deviated substantially from the original purpose which precipitated people to subscribe. He wrote, 'this scheme is not only distasteful to some of the most generous subscribers to the fund, but is also greatly resented by those in a position to voice the feelings of the relatives of our dead heroes and their surviving comrades, in whose interest the fund was originated'.[35] O'Higgins' objections, however, were more closely related to the spaces of commemoration, the seat of government and the narrating of national history. If there was a history of nationhood, there was also a historical geography of national self-determination. From O'Higgins' perspective, the location of a memorial park in Merrion Square would:

give a wrong twist ... to the origins of this state. It would be a falsehood. You have a square here, confronting the seat of the Government of the country ... that any intelligent visitor, not particularly versed in the history of the country, would be entitled to conclude that the

---

[34] Dr W. B. Yeats, *ibid.*, cols. 444–6.
[35] Lord Glenavy, Dáil Éireann, *Official Report*, xix, col. 399, 29 March 1927.

origins of this State were connected with that park and the memorial in that park, were connected with the lives that were lost in the Great War . . . That is not the case. The State has other origins, and because it has other origins I do not wish to see it suggested, in stone or otherwise, that it has that origin.[36]

The connections between the spatial arrangement of sites of memory and the past that they seek to narrate is clearly exposed in O'Higgins' reading of the proposal. While he acknowledged the grief experienced by the relatives of dead soldiers (some of his relatives were killed in the war) and the sacrifice endured by the soldiers themselves, he nevertheless sought to distance the foundations of the Irish state from the conflict in Europe. The origins of the state, in O'Higgins' view, emanated from the period around 1908 when constitutional agitation gathered pace and culminated in the war at home. He stated: 'A revolution was begun in this country in Easter, 1916. That revolution was endorsed by the people in a general election in 1918, and three years afterwards the representatives of the Irish people negotiated a Treaty with the British Government.'[37] The Executive Council would not object to a memorial park at some other site in the city. It was, indeed, the geography that was crucial in this proposal. The symbolic connections which, O'Higgins alleged, would be made by mapping commemorative space alongside parliamentary space underscores the belief that the general population would make the same reading of the spatial connections as the politicians. The view expressed here underscores Lefebvre's more general observation on the significance of space in the conjugation of meaning. He reminds us that 'a spatial code is not simply a means of reading or interpreting space: rather it is a means of living in that space, of understanding it, and of producing it'.[38]

Not all deputies concurred, however, with the Executive Council's position. A variety of arguments were made on the floor of the House to dispute the historical trajectory suggested by the government. Captain William Redmond, a veteran of the Great War, queried the contention that either the subscribers or the promoters of the Bill intended in any way to connect the origins of the state with the Great War. Given that it was a Private Bill and not involving state funds, Redmond suggested that it ought to pass a second reading and be adjudicated on its merit by a Joint Committee of the Dáil and Senate, as laid down by parliamentary procedure. To reject the Bill at this stage, Redmond contended, would be to suggest that those proposing the project were not worthy of receiving a fair hearing as citizens of the state enjoying the same rights as all other citizens. On a more poignantly political note, Redmond observed, that 'it is clear to everyone who has at heart the ultimate re-union of our country that what is needed most to-day is a policy of appeasement and of reconciliation to bring about the effacement of past feuds and to make way, if we can, for a proper and healthy national development'.[39] The government's

---

[36] Mr O'Higgins, *ibid.*, col. 400.   [37] *Ibid.*, col. 403.
[38] Lefebvre, *The production of space*, 47.
[39] Captain Redmond, Dáil Éireann, *Official Report*, xix, col. 407.

position, in his view, would be likely to cause offence to those who wished to offer the city a place of commemoration. In addition, he endorsed the proposal on the grounds that it would provide an important facility for the poorer children of the city. With respect to possible damage and trouble that might emanate from the holding of Remembrance Day commemorations in the park, Redmond suggested that a clause could be inserted into the Bill providing that no such assembly could take place. Finally, Redmond questioned the logic of O'Higgins' argument about location and the inferences that might be drawn about the origins of the state. Taking by way of illustration the memorial arch at the entrance of St Stephen's Green, which commemorates those Dublin Fusiliers who died in the Boer War, Redmond fumed that:

There is a monument in the very centre of our city, probably the most prominent place in the commercial and residential quarters of Dublin which nobody in their senses would suggest had the approval even of a very small percentage of the Irish people. Does anybody coming to Dublin and looking at that gate and at that memorial think that the action of these men had anything to do either with our history in the past or with our future?[40]

Redmond's contribution to the debate highlights the confused meanings associated with public memorials. Private grief and public acknowledgement constantly conflicted with each other. If public statuary exercised no influence in the constitution of a collective consciousness, it was unclear why states and other organisations expended so much financial and ideological energy debating and planning it. Redmond was aware, however, of the potential offence that a rejection of the Bill could lead to both at home and abroad, especially in Britain.

Deputy Shaw – chairman of the Advisory British War Pensions Committees for the western counties – informed the Dáil that he was instructed to oppose the Bill by representatives of ex-servicemen. While the issue of disturbances at the 11 November commemorations did not hold significance for Shaw, 'as the small section of the community who object to Ireland honouring her dead are unworthy of any notice, and have only earned contempt by their cowardly interference with the celebrations',[41] the more important objection related to the expenditure of the money raised. In particular, Shaw suggested that ex-servicemen would prefer to have the money spent on establishing industries, providing employment or housing for veterans and that a sum of £10,000 be spent erecting a memorial perhaps in the Phoenix Park. By contrast, Sir James Craig, a commissioner for the square, wholly endorsed the project emphasising the social and health benefits the square would provide for the inhabitants of the poorer areas of the inner city adjacent to the square. By transferring the square from private to public ownership, the project would enhance the quality of life of the city's children, in particular. The commissioners, he went on to assert, would be within their rights to ban any assembly taking place in the square and thus avoid any possibility of confrontation

---

[40] *Ibid.*, col. 410.   [41] Mr Shaw, *ibid.*, col. 414.

between citizens of the state. Appealing to the Executive Council to alter its position 'because of the public good that would ensue',[42] Deputy Craig gave the proposal his full endorsement. Major Cooper, another veteran, who took exception to the arguments forwarded by O'Higgins, also supported the Bill. Reading the origins of the Irish state rather differently, Major Cooper suggested that the Truce of 1921 was partly procured through the influence of the government of the United States on British opinion. He argued, that 'the United States of America would have been very much less in favour of a peaceful settlement in Ireland had it not been for the services rendered by the Irish soldiers in the Great War, side by side with the soldiers of the United States'.[43] Commenting on the lack of generosity exposed by the ministers of government in their attitude to this proposal, Cooper pointed out that no deputy opposed the allocation of a sum in government estimates the previous year to fund a memorial in Glasnevin to the Easter Rising, 1916. No opposition was launched because 'we did not want to revive old wrangles and old quarrels, because we think the country ought to look to the future while honouring those who died in the past'.[44]

In the early years of the Irish Free State, however, it was difficult to so easily dispose of the past. Deputy Byrne also questioned the government's interpretation of history, and the links between the Great War and Irish independence. For Byrne, '... brave and great Irishmen gave up their lives and fought in France in the belief that they were fighting for Ireland, and I am satisfied, when the Treaty was signed, that the work of these men and the sacrifices made by them, were not forgotten, and in no small way led up to the Treaty'.[45] This comment mirrors the remarks made in Ulster which emphasised that Ulster's sacrifice in the Great War might be rewarded by a continued place within the union. In addition, Byrne pointed out that some of the new national army formed in Ireland after independence recruited ex-servicemen who served in France. The links between the First World War and Irish statehood could not be so easily disentangled for this deputy. Deputy Good similarly took exception to the minister's speech. He claimed that it gave a political significance to the memorial no matter where it was located and would prevent the hope 'that this Memorial which all parties representing all political creeds had subscribed to so freely would have been kept clear of politics'.[46] The deputy was of the view that, on the grounds of parliamentary principle, the Bill should pass its second reading and be sent to a Select Committee for judgement.

By contrast, Deputy Doyle suggested that there was much opposition to the proposal both inside and outside the Dáil chambers. He claimed that the Bill was intrinsically contentious for the population at large and that it could become the site for diverse political opinions to enter into open conflict, as had happened previously in the city during Armistice Day rituals. While Deputy Doyle did not elucidate the precise lineaments of the debate it is quite clear from his observations that a decade after the war Ireland continued to have difficulty in positioning the

[42] Sir James Craig, *ibid.*, col. 419.   [43] Major Cooper, *ibid.*, col. 421–2.   [44] *Ibid.*, col. 423.
[45] Mr A. Byrne, *ibid.*, col. 425.   [46] Mr Good, *ibid.*, col. 427.

war in its own political and cultural history. Focusing on the benefits of a public park for children of the city, Doyle suggested, would be to deflect attention away from the real issue related to the articulation of public memory in the early years of the Irish Free State by the Merrion Square project.

Deputy Lyons, on rather different grounds, opposed the Bill because his constituents, representing a branch of the ex-servicemen's association, thought it an extravagant use of money to spend on a memorial. While they favoured the erection of a cheaper one in Phoenix Park, they felt that the money could be better utilised to serve the families of dead soldiers. The tension between serving the living and remembering the dead is again here expressed. The necessity to balance the needs of the survivors with the public acknowledgement of a debt to the deceased recurs in this parliamentary debate. In a closing statement, O'Higgins repeated his government's opposition to the Bill using an organic metaphor to reiterate his case: 'A tree can but grow from its roots. If you try to substitute others you have a poor tree. This state has particular origins, and particular roots, and we should not suggest either to ourselves or to people coming here amongst us that it has any other roots.'[47] For O'Higgins, those roots were to be found in the political events at home rather than in the role of Irish soldiers in the war in Europe. His position suggests the impossibility of the state having multiple roots that emanated from a variety of political and cultural processes. Nevertheless, he did not seek to underestimate the necessity to memorialise the dead, but 'to distinguish very clearly between commemoration of the dead and glorification of the living'.[48] The approval of the Merrion Square site would be likely to detract from the solemnity and reverence due on 11 November by confusing the public about the role of the First World War in the constitution of the Irish parliament. This confusion would be entirely unnecessary, according to O'Higgins, if another site could be found for the memorial. The Bill was defeated (13 for, 40 against).

This was the first public debate on the role of the First World War in Irish history. The substance of the debate reveals three separate lines of argument developed from radically different political positions. First, there was the government's concern over the geographical proximity of the park to the parliament and the connections that could be made between the two. As Sherman has observed: 'Prompting memories as discrete images, but depriving us of the narrative fabric we weave in our own lives, places stand for both the continuity and the disjunction between past and present.'[49] The government feared that by placing the memorial opposite parliamentary buildings this paradox might be exposed. Second, there was concern over the role of the park as a site for civic improvement for the citizens of the inner city. Veterans and representatives of the ex-servicemen's organisations seemed to want to distance themselves from the park proposal because of its apparent civic rather than commemorative role within the city. One senses that they feared a dilution in the meaning that would be attached to the park as a site for entertainment rather

[47] Mr O'Higgins, *ibid.*, col. 433.   [48] *Ibid.*, col. 433.
[49] Sherman, *The construction of memory*, 215.

than commemoration. Their fears reflect the observation made by Halbwachs, that 'There is no universal memory. Every collective memory requires the support of a group delimited in space and time.'[50] In the Irish Free State of the 1920s support was at best ambivalent. Third, a group of senators and parliamentary deputies questioned the overall value of expending so much money on a park while veterans were facing hardship and unemployment. Their concern centred on achieving a balance between the needs of the living veteran and society's debt to the dead. The tension between these two objectives is clear from the substance of the debate. These three underlying themes therefore were to dominate the debate about an Irish national war memorial and whilst historians have emphasised the first of these concerns it is clear that consensus was far from achievable on the two other issues when one examines the substance of the debate. While there was parliamentary debate then about the status and siting of a national memorial, individual communities were busily planning their own memorials.

Church memorials

In 1920 at St George's parish church in Dublin the Protestant archbishop of Dublin unveiled a memorial window and tablet dedicated to the 84 men (from a total of 450 volunteers) of the parish killed in the war (see Figure 21). Costing a total of £350, the memorial was placed in the east side of the north gallery of the church and 1,200 people attended the ceremony. The memorial window comprised a lower portion representing a dying soldier among the ruins of churches and homes – a landscape typical of the devastated areas of the Western Front – with shattered guns and war debris strewn around, and a central panel glorified the figure of the Saviour ascending among the angels: a scroll bore the following inscription: 'Greater love hath no man than this, that a man lay down his life for his friends.' Two other angels bearing the chalice of victory and the triumphal crown accompany him.

The secular and the sacred are inextricably linked in this memorial tablet. The border of the window comprised a roll of honour, formed by a succession of laurel wreaths each enclosing a group of names of the 84 killed. The ceremony was accompanied by Chopin's funeral march, performed by the North Dublin Choral Society, the *Last Post* was played and the National Anthem sung at the end of the service. In the porch of the church was a brass tablet containing the names and regiments of each of the 84 men and this was totally funded by a Mr H. Darker in memory of his son killed in service. At St George's church then both the congregation and an individual contributed to the creation of a memorial space within 'sacred' space.[51]

Similarly in 1920 at St Patrick's church in Dalkey, Co. Dublin a memorial tablet was dedicated and unveiled by Reverend Collins, dean of Belfast. Situated within the chancel, the memorial was made up of panels divided by green Connemara

[50] Halbwachs, *On collective memory*, 84.   [51] *Irish Times*, 29 March 1920.

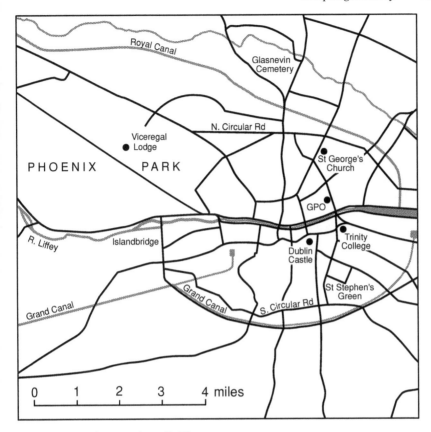

Figure 21 Location map: inner Dublin

marble columns and capped with polished marble. One large central panel carved of white marble contains the tablet bearing the names of the sixteen members of the congregation who died in the war.[52] In Cork, the rector of Christ Church informed the congregation that the Baptistery had been renovated as a war memorial and a handsome stained glass window inserted, accompanied by white marble tablets listing the dead.[53] At the Presbyterian church in Bray two memorial windows were unveiled on Easter Sunday in 1925 to a large congregation which included ex-servicemen wearing their decorations and singing the anthem 'I know that my Redeemer Liveth'. Reverend Simms in his sermon spoke of being near Ypres in 1915 chatting to an officer about the great changes that the war would bring about and the new world that would arise out of the old. The officer was killed three days later. Simms wondered 'What a disappointed man would that officer be with what they saw today. They [the soldiers] understood the meaning of sacrifice. They knew

[52] *Irish Builder*, 17 July 1920, 470.   [53] *Irish Builder*, 24 April 1920.

that it was because they themselves were content to die that the British nation was enabled to bring back victory.'[54] By the mid-1920s, however, Ireland's constitutional links with the union had been severed and Easter Sunday was in some circles a celebratory occasion for the sacrifice of those who died for the establishment of independence. The commemorative calendar conjugated historical significance in different ways.

Church memorials could be planned and executed by local congregations and did not need the sanction of the state or veterans' organisations. In that sense they more explicitly served religious communities, specifically Protestant ones. Catholic churches rarely included the iconography of civil society within its spaces. The separation of the secular world from the sacred one of the church interior ensured that national flags, military memorabilia or explicit representations of war could not be incorporated into the visual lexicon of Catholic worship. While iconographic representations of religious texts and sacred narratives saturate the space of many Catholic churches, the inclusion of icons of remembrance to war would blur the clear distinctions between the worlds of the sacred and the profane. Consequently, church memorials and the services surrounding them formed a far greater part of a Protestant calendar of worship than the Catholic one and it enabled communities to remember their dead locally without interference from higher religious or secular authorities.

Mapping memory in the public sphere

Throughout the 1920s, there were efforts made in towns and villages in the Irish Free State and in Northern Ireland to create landscapes of remembrance. While Protestant churches in particular were quick to establish memorial tablets inside their premises, public monuments took longer to execute. This in part was the result of the variety of ex-servicemen's associations in existence, including the Irish Nationalist Veterans' Association (active in the early 1920s); Comrades of the Great War (later to be called the Legion of Irish Ex-servicemen) and the British Legion (the latter two were affiliated in the mid-1920s).[55] Raising funds, acquiring public sites and employing designers all contributed to the delay, and as the years passed perhaps the urgency to remember had receded among the general populace although clearly not among veterans or the families of the dead.

One of the earliest memorials to be built was the one in Bray, Co. Wicklow, unveiled in 1920. Under the chairmanship of Viscount Powerscourt, the war memorial committee rapidly raised funds and proposed a Celtic cross to be located on a plot of land opposite the Princess Patricia hospital which had been donated by the railway company.[56] Built of Tullamore limestone and designed by the architect Sir Thomas Deane, the cross was 17 feet high, with a pedestal and eight bronze

---

[54] *Irish Times*, 13 April 1925.   [55] Jeffery, 'The Great War and modern Irish memory', 148.
[56] *Irish Builder*, 22 February 1919. See also Leonard, 'Lest we forget', 59–67.

Figure 22 Memorial in Bray, Co. Wicklow

panels. Six of the panels contained the names of local soldiers killed in the war with the inscription, 'This cross is erected by the people of Bray in loving and grateful memory of the brave sailors, soldiers and airmen who gave their lives for their country in the Great War' (Figure 22). The Celtic cross had experienced something of a revival for gravestones and funerary commemorative monuments since the mid-nineteenth century. Copied from the ancient high crosses and made popular through the publication of a book on sculptured crosses by Henry O'Neill in 1857,[57] the Celtic cross motif had already been used in the centenary

---

[57] J. Sheehy, *The rediscovery of Ireland's past: the Celtic revival 1830–1930* (London, 1980).

commemorations of the 1798 rebellion.[58] Although a religious icon, the pre-Reformation origins of the cross's design facilitated its use by both Catholic and Protestants.

St Patrick's Day, 17 March 1925, was selected for the unveiling of the Cork City war memorial. Having been granted planning permission by Cork Corporation to erect a monument along the South Mall, one of the main thoroughfares in the city centre, the Cork Independent Ex-Servicemen Association commissioned the design and organised the unveiling of the monument. On the morning of the unveiling, Catholic veterans attended services in St Mary's Cathedral while Protestant veterans attended services at St Finbarr's Cathedral before joining together to parade to the monument at 11 o'clock in the morning. The British Legion also took part in the ceremony. General Harrison and the executive committee responsible led the parade. Men wore military decorations, women held wreaths and some children wore the medals their fathers had been awarded. Six bands took part in the procession, which reached the memorial shortly after 2.00p.m. The ceremony was presided over by Gerald Byrne, chairman of the Cork Ex-Servicemens' Association, who in his opening address stated that they were assembled to 'unveil a monument to their comrades who fell on the different fronts fighting for the freedom of small nations'.[59] The meaning of the war in Cork could be legitimated through an appeal to the defence of small nations. General Harrison, appointed to unveil the monument, focused on the need for government support to veterans and their families. According to Harrison, they deserved state funding: he finished his speech with the phrase 'God Save the King and God Save Ireland', a diplomatic tactic which avoided alienating those with nationalist or unionist political sympathies. The monument itself, however, was shrouded in a Union Jack and the Reveille was played when it was removed. With independence achieved perhaps the Union Jack had lost its political resonance. Mr Egan TD who addressed the audience expressed his privilege at attending the unveiling ceremony. He stated:

it was their duty not to forget the brave deeds of men who went out and died on behalf of the small nations of the world . . . When the question was raised in connection with the erection of that monument some friends and himself had stepped into the breach because they saw no reason why Irishmen who died abroad could not be remembered by the people whom they served, and a memorial put up to their memory in their native city.[60]

He also encouraged the government to cater for the needs of veterans and their families. Thus, while Egan emphasised the purpose of the war in relation to the protection of small nations, he saw no contradiction between commemoration and the new Irish state.

Following the same theme, John Horgan's speech also underscored the role of the sacrifice of men on the Western Front to Irish political ideals. He commented

[58] Johnson, 'Sculpting heroic histories', 78–93.   [59] *Cork Examiner*, 18 March 1925.   [60] *Ibid.*

that these men 'did as much as any people in Ireland in order that the people of the country might be free. He remembered well when the rallying cry for the freedom of small nations rang out, how the men of Cork responded in order to achieve such an object and particularly to prove the right of their own land to win its liberty.'[61] In Cork, then, the vocabulary of the Great War existed comfortably alongside the independence movement. Unlike the ruminations about the National War Memorial Park in Dublin, the speeches on this occasion emphasised the connections between the establishment of the Irish Free State and participation in the war. These ideological links were not regarded as mutually exclusive or contradictory. National newspapers reported the unveiling as a successful event and one that emphasised the unity of purpose between all those who participated in the war.

The iconography of the memorial itself differed from many others around the country (Figure 23). It comprised three granite squares forming a pedestal, which carried a bas-relief of a soldier, in military uniform, with his head bowed. The carved soldier appears only on the front side of the sculpture. The entire monument was 20 feet in height and contained the inscription 'Lest we forget'. The plinth stated that the memorial had been erected by public subscription under the auspices of the Cork independent ex-servicemen's club 'in memory of their comrades who fell in the Great War fighting for the freedom of small nations 1914–18'. Although the design was simple, it did not draw on the imagery of the Celtic Revival which characterised much commemorative statuary in Ireland. Figurative but unheroic, the statue underscores the isolation and grief of the individual common soldier.

In the border county of Longford, about 80 miles north-west of Dublin, plans too were afoot to mark the contribution of that county to the war effort. 'It is to be distinctly understood that no flags are to be displayed or carried in the ranks at any time during the day.'[62] These were the instructions issued by the War Memorial Committee for the parade accompanying the unveiling of the war memorial in Longford town in 1925. Longford town council had granted a permanent site in the Market Square for the erection of a memorial to the soldiers of the Great War, organised under the stewardship of the earl of Granard and his Co. Longford committee. As one of the earliest public monuments to be unveiled in the Irish Free State, many of the speeches emphasised that Longford might set a precedent for other towns and cities.[63] The unveiling ceremony involved a full parade of ex-servicemen who gathered at Church St at 1.30p.m. The procession included a wide variety of bands including the band of the British Legion from Dublin and the Irish National Foresters Brass and Reed band, followed by the organising committee, officers, dignitaries and guests, ex-soldiers and comrades of the Great War. The parade marched from Church St to the Market Square, where the memorial was unveiled and continued another circuit of the town back to the memorial where Sir William Hickie took the salute before the parade was dismissed (Figure 24).

[61] *Ibid.*   [62] *Longford Leader*, 21 August 1925.   [63] *Longford Leader*, 22 August 1925.

Figure 23 Cork City memorial

Most men did not have uniforms but were advised to wear ordinary civilian attire so as to be as alike as possible irrespective of rank. It was reported that in the parade through the streets of the town there were many clergymen and a number of distinguished officers. It is estimated that there were about 1,000 spectators, who travelled from around the county to the ceremony.[64] About 300 Longford men had been killed in the war and the construction of a public statue was intended as a tribute to them. The parade was accompanied by the playing of the 'Dead March' by bands from Dublin and Sligo. At Market Square, where a platform draped with Union Jacks had been erected, Sir William Hickie unveiled the memorial, also

[64] *Ibid.*

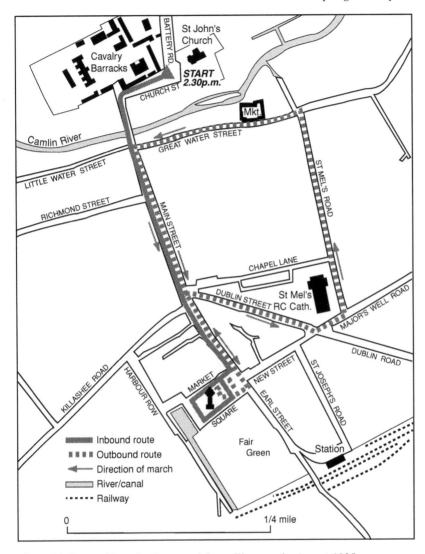

Figure 24 Route of Longford's memorial unveiling parade, August 1925

covered with a Union Jack. The 'Last Post' was played, followed by a 2 minute silence. Although not Armistice Day, the ceremony adopted much the same format, and once the silence had been observed 'God Save the King' was played. Market Square itself was a riot of colour combining flags of the Union Jack with other allied flags.[65] Major Bryan Cooper TD, had the following to say:

[65] *Longford Leader*, 29 August 1925.

Memorials such as this are to be found all over France, Italy, and Great Britain, but in the Free State they are very rare. Longford has set the example for the rest of the country, which, I hope, will be widely followed ... Acts of commemoration such as this are the last and most loving tribute that we can pay to our comrades.[66]

The fact that the iconography surrounding the ceremony might seem to conflict with the newly established independence of Ireland did not feature in Cooper's analysis. Lord Granard presided over the unveiling meeting. He paid a glowing tribute to the soldiers from Longford and connected their history to the broader history of Europe. Hickie similarly linked the battlefields of European history with Irish blood commenting that 'whenever the British Army took the field, there also, was the Irish soldier to bear them company and to take his part'.[67] He also hoped that the prominent site afforded by the town council to the Longford memorial would serve as an example to the legislature in Dublin where 'the same broadminded spirit, the same desire to do honour to the men ... may dominate their counsels'.[68]

Lord Longford's address also stressed Ireland's military prowess observing that 'For centuries Irishmen have been the first soldiers of Europe.'[69] Connecting Ireland to a European military tradition may have had some rhetorical resonance for the listeners and served to distance soldiers of the First World War from some unofficial armies present in Ireland. Major Bryan Cooper outlined the motivations of Irish volunteer soldiers: 'They are the men, whom, of their own free will, left home and fireside, and all that makes life dear, to sacrifice their lives for a great cause. They and their deeds are immortal.'[70] But these deeds could be made immortal through the act of memorialisation. The memorial itself, drawing from the iconography of the Celtic Revival comprised a plain Celtic cross 20 feet high, carved from locally quarried limestone. While Celtic crosses were popular commemorative icons in the Irish Free State, they found far less appeal in Northern Ireland. Jeffery surmises that this may be due to 'Northern Protestants' traditional unease with the cross as a religious symbol'.[71] In an overview of British war memorial iconography, Moriarty notes that the use of crosses and other religious icons was widespread and their unveiling ceremonies were surrogate funeral services, where the cross represented sacrifice and the hope of resurrection for the bereaved families. While the bodies of the dead were buried abroad, the local memorial became, in some respects, the shrine of remembrance and the Church played a significant role throughout Britain in the execution of public remembrance. Celtic crosses proved extremely popular designs for many memorials. Indeed Moriarty claims that 'A cross is by far the most common type of First World War Memorial in Britain',[72] and the relative absence of crosses in a Northern Irish context is

[66] *Ibid.*   [67] *Irish Times*, 28 August 1925.   [68] *Ibid.*   [69] *Ibid.*
[70] *Ibid.*   [71] Jeffery, 'The Great War in modern Irish memory', 147.
[72] C. Moriarty, 'Christian iconography and First World War memorials', *Imperial War Museum Review*, 6 (1990), 69.

thus somewhat surprising. The local religious context may have been an important consideration in the selection of memorial icons. At the memorial in Longford, several wreaths laid were made of Flanders poppies, palms and laurel leaves, with a heart-shaped wreath delivered by the Longford ex-servicemen. The poppy wreaths were to be preserved in the Legion's club premises to be re-laid at the memorial on the 7th Armistice Day.[73] The day in Longford was adjudicated as having been a successful event and one that gave a permanent place to Longford men killed in the war.

The siting of war memorials was important in any context. In Britain, they were generally located on church grounds or in a significant civic space, where the latter 'provided a sacred symbol ... yet were also a highly public statement of a district's involvement with the war'.[74] The transformation of public, secular space into sacred sites of mourning represented an interesting admixture of civic and spiritual responsibility, political and cultural cooperation. In Ireland, the choice of site frequently proved problematic, as local authorities had to confront diverse ideological opinions within their council chambers. The scripting of the war on the Irish landscape continued to compete with sacrifices made at home and their recognition in public space. In 1928 in Sligo, for instance, there was a heated exchange between members of the corporation and the local branch of the Legion regarding the allocation of a site for a memorial. Although Sligo Corporation had approved a site in the town centre it transpired at the last minute (a week before the proposed unveiling) that the site stood above a main town drain and the Legion requested that a new site, directly beside the old one, be approved immediately by the local authority. The mayor of Sligo argued that the old corporation had granted permission, but that new members would have to be fully informed and a hasty decision could not be made. He also commented that it was not a matter of great urgency as the war had ended over ten years ago.[75] Nevertheless, a special meeting of the corporation convened the following day involving the mayor, alderman, eight councillors and two representatives of the Legion's memorial committee. Councillor Kelly proposed that an alternative site be granted with the rider that he hoped that

they [Legion] are not going to make any propaganda out of the unveiling of the memorial on Sunday. We don't want Major Cooper or any of those people making propaganda out of it ... We don't want bands playing 'God Save the King' or 'God save the country' because God gave me brains to think and I have come to the conclusion that if there were less Kings and Queens, Coopers and Hickies there would be less men dead and there would be no need for this memorial.[76]

Councillor McMorrow seconded the motion but was of the opinion that there was not any British imperialism about it, while Councillor Joyce felt that the band could play whatever music it desired.

[73] *Longford Leader*, 29 August 1925.   [74] Moriarty, 'Christian iconography', 70.
[75] *Sligo Independent*, 29 October 1928.   [76] *Ibid.*

In the meantime it emerged that Mr Gormon (president of the local branch of the British Legion and a member of the memorial committee) had consented to give a plot of his land just outside the town at Edenville for the memorial. Councillor Fowley commented that 'It will be a place where in the future we can hold our Services of Remembrance without inconveniencing or offending anyone.'[77] This prompted Councillor Dorrig to ask if 'you [are] going to put it where no one will see it?'[78] The invisibility of the site from the public gaze underscores the equivocal support such projects received in certain quarters. Nevertheless, now that a new site had been acquired independent of the local authority, the Corporation was anxious that it not be seen as obstructing the wishes of the Legion. Councillor Morrow made the case that he trusted that 'the relatives of the Fallen will not think that we are in any way responsible for not having the memorial at the Ulster Bank [original site]', and Councillor Depew observed that 'The Corporation has got a very bad name over this matter, and we must ask the Press to make it clear that we were quite willing to grant an alternative site.'[79] Thus although certain council members sought to make life difficult for the Legion, they were simultaneously anxious to avoid public ridicule in their handling of the issue.

The memorial was unveiled on the new site the following Sunday by General Hickie (senator). The local newspaper situated Ireland's role in the war in a larger geopolitical context where Sligo men contributed to the curbing of German and Central European militarism. According to the *Sligo Independent*:

There can be no doubt about it, the foul feet of the hideous monster of militarism would have stamped over our land . . . Some do not believe in spending money on memorials of this kind . . . it is right, fitting and natural that there should be something tangible, something for everyone to see . . . as a symbol of the stout and hardy Sligomen whose supreme sacrifice should not be forgotten while time endures . . . Never in the history of the world was there a juster cause.[80]

The First World War was understood, then, as an exceptional period in the history of humanity where the moral imperative to defend against the beastliness of German expansionism and cruelty was unequivocal. The moral equation could easily be calculated and the righteousness of the men of Ireland unquestioned. While this discourse was popular in the years during the war, particularly in the recruitment propaganda of the day, it is somewhat surprising to see such fervent anti-German expression at a commemorative ceremony ten years after the war. Perhaps the tone of the newspaper report can be accounted for as a rejoinder to those who questioned the motives behind the war and Irish men's role within it.

In contrast to the ceremony in Longford, the memorial was draped in purple and the foreground of the site was strung with Marine Signal flags. A large crowd[81]

---

[77] *Ibid.*   [78] *Ibid.*   [79] *Ibid.*   [80] *Sligo Independent*, 27 October 1928.

[81] The *Irish Independent* claimed that several hundred ex-servicemen attended. There was a detachment of the Boy Scouts from Sligo Grammar School, the British Legion HQ's band, Sligo town band and a large public attendance. Apologies were read from the bishop of Elphin, General Mahon, Major

attended the unveiling dedicated to the more than 400 Sligo men who died in the conflict, of 'all creeds and classes mingled in true democratic fashion in the throng in front of the memorial'.[82] The band played the 'Dirge for the Dead' and the 'Funeral March' marrying the ecclesiastical with the secular. After a short speech congratulating the men of Sligo for defending against the pan-Germanic aspirations of the Central Powers, Hickie unveiled the memorial – a Celtic cross. Made of Irish limestone, the inscription read 'In Glorious Memory of the Men of the Town and County of Sligo, who gave their lives in the Great War, 1914–1918.' The inscription had a distinctly local complexion to it, bringing the war home to the locals. After the unveiling the 'Last Post' and 'Reveille' were played, wreaths were laid and the crowd dispersed. It was described in the newspaper 'almost as if the whole ceremony were a symbol of life and death'.[83] Life was expressed through the survivors and their symbols of survival – the hand-made wreaths – death through the stone structure representing those who never arrived home. Each local ceremony would serve to bring back remembrance of the ordinary but equally worthy soldier and would serve to remind the population of the price of peace. As Winter has suggested, traditional forms of remembrance were regularly more effective for bereaving families than abstract and cynical forms of representation. Traditional languages and spaces of commemoration had the potential to heal where modernist abstraction might alienate.[84]

In Ulster the creation of civic landscapes of remembrance to the fallen of the Great War aroused, in some respects, far less controversy. The significance of the Battle of the Somme embraced through Orange parades immediately after the war was supplemented by the erection of memorials to the war. As Foster has reminded us, the Somme represented 'an archetypal event in loyalist psycho-history'.[85] One of the first acts of public memorialisation therefore was the erection of a monument at the site of battle, to the efforts of Ulster in the war.[86] While the Somme battlefield is an important location for many of the armies of Britain's empire, including the Canadian, South African, New Zealand and Australian armies, the first memorial to be erected in this space was the Ulster Tower.

Paid for by public subscription, and planned since 1919, the tower was dedicated in 1921 and is located just west of the rebuilt village of Thiepval (Figure 25). The site represents the portion of the Allied front line occupied by the 36th Ulster Division on the morning of 1 July 1916, and this fact reinforces the significance of the physical site in the construction of meaning. The tower is an exact replica of Helen's Tower on the Clandeboye estate, Co. Down, a Scottish baronial tower erected in 1861. The monument was chosen by James Craig (later prime minister of Northern Ireland) and Colonel Wilfrid Spender (later secretary to the Cabinet

Cooper TD, Major Tynan and Reverend Fr Stafford (a chaplain to the 10th Irish Division). *Irish Independent*, 22 October 1928.

[82] *Ibid.*   [83] *Ibid.*   [84] Winter, *Sites of memory, sites of mourning.*

[85] J. W. Foster, 'Imagining the *Titanic*' in E. Patten, ed., *Returning to ourselves* (Belfast, 1995), 334.

[86] Ulster division memorial, October–December 1919, PRO WO 32/5868, nos. 6, 10, 10A and 11.

Figure 25   Ulster Tower, Thiepval

under Craig's government). Both had served in the 36th Ulster Division. Some of the troops of that division were trained on the estate in 1914 and 1915. The naming of the tower emphasised its regional rather than its national character and the dedication at the entrance of the tower reads 'Memorial to the 36th (Ulster) Division and to other men of Ulster who served in the Great War 1914–1918'. While there is no reference to religious identity, the regional reference, in some respects, can be seen as trope for political and religious allegiances. As Sherman has noted, 'the sites chosen for monuments established a concrete set of relationships between commemoration and the life of the community'.[87] The choice of

---

[87] Sherman, *The construction of memory*, 217.

Thiepval and the emphasis given to the 36th Ulster Division at the site reinforced the significance of the war for Ulster's Protestant community. While there were class and denominational differences within that community, the war could be remembered as a collective enterprise whose ultimate consolation was the creation of the Northern Ireland state. In that sense 'the displacement and appropriation of individual mourning by collective tribute'[88] was given added meaning. Although other memorials to Irish divisions were erected in France and Belgium in the 1920s and 1930s, the Ulster Tower was the largest both in terms of scale and symbolism.

At home, also, Ulster set about erecting memorials to the dead, some within the spaces of Protestant churches and others in town squares. One of the most significant church memorials was erected in St Anne's Church of Ireland cathedral, in Belfast, where the west portals were dedicated 'to the men of Ulster who fell in the Great War'. With respect to local memorials it has been observed that 'communities chose signs that represented their sense of themselves, of what distinguished them from others'.[89] In an Ulster context religion was the most significant marker of difference. The placement of district memorials on the grounds of Protestant churches, as happened in Kilkee, Co. Down, inevitably conveyed a sense of distance from acts of commemoration among Ulster Catholics.[90] By treating the war as a signifier of Ulster's loyalty and by rewarding the sacrifice with partition, it was unlikely that commemorations to the war would hold much appeal to northern Catholics in the early years of the new state.

The placing of Belfast's main memorial within the grounds of the Protestant-dominated council offices at City Hall ironically was to make the connections between the war and politics in ways that were being wholly resisted in Dublin (see Figure 20). Sherman reminds us that 'Monument sites prompted strong reactions, moreover, because they entailed a kind of geographical superposition of memories: memories of individuals had to share mental space with the memories attached and attributed to places.'[91] As the epicentre of local government in Belfast, the City Hall may have been an uninviting venue for commemoration among the city's Catholics. But the ritual of remembrance was pronounced in that city, with the Armistice Day programme of 1926 listing thirty-nine representative bodies laying wreaths at the cenotaph in City Hall. These included the lord mayor, prime minister and governor of Northern Ireland.[92] In contrast to the Irish Free State, the narrative of remembrance embraced the highest echelons of Northern Ireland's Protestant political elite. The unveiling of the cenotaph in 1929, according to Jeffery, was 'almost exclusively a Protestant affair'.[93] The 16th Irish Division was omitted from the invitation list, although the Italian Fascist Party was included. The following year, however, the 16th Irish Division was invited. The choice of a cenotaph as the symbolic representation of loss perhaps reflected a desire to replicate the pattern

---

[88] *Ibid.*, 218.   [89] *Ibid.*, 217.   [90] Jeffery, *Ireland and the Great War*.
[91] Sherman, *The construction of memory*, 218.   [92] Gregory, *The silence of memory*.
[93] Jeffery, *Ireland and the Great War*.

in London, although the anonymity of identity implied through use of an empty tomb may not have had the intended impact of inclusivity in a heavily religiously segregated city.[94]

Overall, in the decade following the armistice, Northern Ireland created a series of memorial spaces to the war that inscribed it on to the historiographical and memorial record. Whilst there were some instances of inclusive rituals of remembrance, the narrative undergirding the overall exercise tended to highlight and reinforce the exceptionality of Ulster within the island and replicate the divisions found in a pre-war Irish context. The fact that the political boundaries of the island had changed since the war's end added weight to the battle of ideologies that had characterised the earlier decades. But while Ulster commemorated, the Irish Free State continued to debate the merits of a national memorial and it is to that question that I now wish to briefly return.

National war memorial revisited

After the rejection of the site at Merrion Square by the government, the Trustees of the Irish National War Memorial proposed the establishment of a monumental arch near the main gate of the Phoenix Park. This proposal was also rejected by the Executive Council.[95] Over the next year a wide variety of proposals were considered by the government including more practical proposals such as a veterans' home, industry modelled on the German home industries for the unemployed, playgrounds in new suburbs and apprenticeship schemes. The notion of a memorial park or a monument came low down in the list of preferences. Nevertheless, by 1929 the government had agreed a memorial park and requested that the Office of Public Works identify a suitable site along the southern banks of the river Liffey. A 25 acre site at Islandbridge[96] (about 3 miles west of the city centre and on the opposite side of the river to the Phoenix Park, see Figure 21) was chosen, and Sir Edwin Lutyens[97] was employed in 1930 as chief architect to design a memorial park there. The park comprised a central Cross of Sacrifice 30 feet high (Figure 26), a Stone of Remembrance with the inscription 'Their name liveth for evermore,' flanked by two fountains in sunken rose gardens (Figure 27: 4,000 roses planted), and four pavilions connected by pergolas containing the *Memorial Records* would

---

[94] The cenotaph was designed by Sir Alfred Brumwell Thomas. Made of Portland stone, the cenotaph stands in front of a curved collonade of paired plain shafts. See P. Larmour, *Belfast: an illustrated architectural guide* (Belfast, 1987).

[95] J. Leonard, 'Lest we forget', in D. Fitzpatrick, *Ireland and the First World War* (Dublin, 1986), 59–67.

[96] 'Proposal outlined at meeting of Cabinet by Mr M. J. Byrne, principal architect, Board of Works', 29 October 1929 (NAI DT S.4156A).

[97] Sir Edwin Lutyens was one of the foremost British architects of the period, renowned for his imperial projects in India, as well as his memorial work in Britain. See R. Gradidge, *Edwin Lutyens: architect laureate* (London, 1981); R. G. Irving, *Indian summer: Lutyens, Baker and imperial Delhi* (London, 1981).

Figure 26 Cross of Sacrifice, National War Memorial, Dublin

complete the design. Rising ground behind the Cross of Sacrifice was terraced and pierced by a broad flight of granite steps flanked by two walls with the inscription (one in English and the other in Irish): 'To the memory of the 49,400 Irishmen who gave their lives in the Great War 1914–18.'

Work began on the site in late 1931 and was virtually completed by 1937 and the park was handed over to the Commissioners of Public Works in 1938. Workmen for the project consisted of ex-servicemen of the British army and ex-servicemen of the Irish army (50 per cent).[98] Classical in conception, the park was a very impressive

[98] British Legion Annual, *Irish National War Memorial* (Dublin, 1941).

Figure 27 National War Memorial Gardens, Dublin

combination of orthodox religious symbolism and secular architecture, with a variety of rose gardens and trees planted around the periphery. The scale of the enterprise and the cost of the park (approximately £100,000 of which £56,000 was contributed by the Irish National War Memorial fund) stands as testament to the government's commitment to establishing a long-lasting memorial to the war. Opening the memorial park officially did prove problematical, with the government initially agreeing to open it in 1938 on the condition that the British Legion agree that no Union or regimental flags would be flown and that the ceremony be confined to ex-servicemen from the Irish republic. In the following year, promises were again made by Eamon deValera to attend the opening ceremony but the imminence of war in Europe and the prospect of conscription in Northern Ireland led to the indefinite postponement of the ceremony.[99] The British Legion, however, did hold its Armistice Day ceremonies in the Park for the following thirty years, not always without controversy. In 1945, for instance, the Armistice parade from Smithfield to the memorial park was banned by the police authorities, which according to the British Legion only served to prompt larger numbers than usual to attend the ceremony in the park: 'registering their indignation at the insult offered to their

[99] For fuller discussion see Leonard, 'Lest we forget'; Jeffery, *Ireland and the Great War*.

noble dead by the heedless and unconstitutional action of the Eire authorities'.[100] It is estimated that 10,000 people attended the ceremonies that year. Some commentators have noted the peripheral location of the park in relation to central Dublin and used this as indicative of the absence of a public commitment to remembrance (see Figure 21). While the merits of the design of the park have been complimented, its geographical location has been regarded as de-centred and as having a de-centring effect on the significance of the war in popular memory and in the official and academic record of the war. Although 3 miles may have been regarded as a relatively far distance in the early decades of the twentieth century, it must also be acknowledged that Glasnevin cemetery, the centre of many republican commemorative occasions, was also that distance away from the city's core. The grand scale of the final design at Islandbridge would have been difficult to accommodate in any place at the city's heart. But perhaps more important than the scale of the ultimate physical space was the symbolic scale of the state's unease with finding a location for the memorial in the political heart of the fledgling state.

Conclusion

In the discussions of Irish memorials space was central to some of the most virulent debates. While national politicians, local councillors, and members of the churches could all agree that some public marking of remembrance to the fallen was desirable, the use of public space for such activity was consistently contested. The anatomy of association between existing sites of memory within cities and towns and new memorials to the war played on the minds of the memory makers, thus making the sculpting of war memory part of a larger process of mapping the nation's history. While in Northern Ireland memorials were common in towns and cities, especially ones with large Protestant populations, for the independent Irish state the scripting of urban space with memorials became a scripting of a sense of national consciousness. If this chapter has focused on the sculpting of public memory through official sites of memory, Chapter 5 shifts focus to forms of representation that were generally more critical in their interpretation of the experience of war and more daring in their expression. In what Hynes refers to as a genre of anti-monuments, which were often 'monuments of loss: loss of values, loss of a sense of order, loss of belief in the words and images which the past had transmitted as valid',[101] these monument makers expressed themselves through painting, poetry, novels, diaries and autobiographies. They too translated the war to the Home Front, through a different medium and often to express a different message and it is with this literary scripting of the war that the next chapter is concerned.

[100] British Legion, *Victory Souvenir* (Dublin, 1946).   [101] Hynes, *A war imagined*, 307.

5

Scripting memory: literary landscapes and the war experience

if Turnage's aim ... was to distil the grotesqueness of war into an aural equivalent of the serial geometrics invented by C. R. W. Nevinson ... he succeeded ...[1]

This quote, from a review of the world premiere of Mark-Anthony Turnage's opera *The Silver Tassie*, which opened in the London Coliseum in February 2000, underlines the continuing imaginative appeal of the First World War as a source of creative energy. While the war was popularly commemorated through monument, memorial and spectacle, the war similarly spawned a vast array of literary works. If social memory found material expression through spaces of commemorative activity, the social imagination was also cultivated through representations of the experience of war in novel, play and verse. The war was mapped and its physical, cultural and psychological spaces made meaningful to popular audiences through these works. Ironically, fiction could at times translate that which documentary accounts found difficult to communicate. The upheavals of 1914 affected every reflective person across Europe: 'Artists, poets, writers, clergymen, historians, philosophers, among others, all participated fully in the human drama being enacted ... Even the introvert Marcel Proust ... was spellbound by the spectacle.'[2] Hynes suggests that literary and artistic accounts of the war created a space for the articulation of an alternative view of the war that deviated from the romantic or heroic expression that appeared in other forms of commemoration. These types of works, he claims, were 'monuments of loss: loss of values, loss of a sense of order, loss of belief in the words and images which the past had transmitted as valid'.[3]

While the First World War provided the subject matter for a mushrooming of literary output, it was also part of a larger set of cultural transformations which were culminating in new forms of expression embraced by the term 'modernism'.

[1] F. Maddocks, 'Turnage scores in injury time', *The Observer*, 20 February 2000, p. 7.
[2] Eksteins, *Rites of spring*, 208–9. [3] Hynes, *A war imagined*, 307.

112

The term is itself controversial and its meaning contested,[4] but, as Tate observes, 'it remains a useful description of writings which were self-consciously avant-garde or attempting to extend the possibilities of literary form in the late nineteenth and early twentieth centuries'.[5] Many interpretations of the cultural history of the war see it as a culmination of the ascent of modernism, where the languages of patriotism and the glory of war, expressed through high diction, were replaced by ironic, abstract and cunningly satirical representations of the war.[6] Others, however, have queried such a strict dichotomisation of literary styles and have emphasised the manner in which traditional modes of representation continued to intersect with more abstract ones in the translation of the war experience.[7]

Perhaps a more profitable point of departure is to suggest that there is a historical geography of literary styles employed in First World War literature. During this period of crisis, approaches to scripting the war were coloured by local circumstances as well as by some of the universal tropes popular both in modernist and non-modernist styles of writing. Emphasis on the uniformity of the war experience for combatant countries may have led to an underestimation of the significance of particularity and geographic specificity to literary narrative. Tropes of war and literary styles travel and circulate across space but they may not do so evenly. It was for this reason that there is the variety of literary depictions of the war. This is particularly well illustrated in Jonathan Vance's study of Canadian war memory. Vance suggests that through bronze and stone, reunion and commemoration, novel and play a mythic version of the war 'became the intellectual property of all Canadians'.[8] While some commentators have claimed that the high diction of the Edwardian world was a spent force, an unsuitable means of communicating the effects of technological warfare, Vance suggests that traditional paradigms persisted and nineteenth-century modes of expression were routinely employed to convey the meaning of the war.[9]

Rather than focusing on the merits of each side of the modern/traditional debate, I want to examine the narrative style employed in specific war writing. In particular, I wish to focus on the structural devices, the metaphorical engagements, the linguistic tropes used which sought to provide an interpretative framework for the popular understanding of the existential as well as social and political spaces of war in Ireland. Themes such as Home Front and war front, secular and

---

[4] For a discussion of the idea of modernism see S. Smith, *The origins of modernism: Eliot, Pound, Yeats and the rhetorics of renewal* (Hemel Hempstead, 1994); P. Nicholls, *Modernisms: a literary guide* (Basingstoke, 1995); M. Levenson, *A genealogy of modernism: a study of English literary doctrine 1908–1922* (Cambridge, 1984).

[5] T. Tate, *Modernism, history and the First World War* (Manchester, 1998), 2.

[6] The most influential exposition of this view is Fussell, *The Great War and modern memory*. A more nuanced version is found in Hynes, *A war imagined*.

[7] This case is powerfully made especially in relation to mourning by Winter, *Sites of memory: sites of mourning*. See also Gregory, *The silence of memory*.

[8] Vance, *Death so noble*, 3.   [9] *Ibid.*

religious motifs, men's and women's role in the war, are narrated in multifarious ways through war fiction. The spaces in which these themes are produced, narrated and received reveals something of the way in which the spaces of war are scripted into popular memory. As Hynes reminds us, 'Stories deal with causality and change, and war-stories tell us processes of war: what happened to the teller, and with what consequences.'[10] In that sense literary interpretations of war are at once personal and collective. The advent of new weapons of technology – aerial bombing, chemical weaponry, the tank – created a different terrain of destruction from which the war would be scripted. The newly advanced technologies of mass warfare in themselves stimulated a range of effects. They accelerated the volume of casualties and the relay of information home via new communication networks. Paradoxically, this made the war seem anonymous, pervasive and dehumanised while at the same time accentuating the personal, individual, located dimensions of grief through the rapid publication of lists of the dead or missing. The modernity of the war itself presented challenges and opportunities for those narrating the experience.

The continued power of First World War novels, poems and plays in stimulating the imagination is revealed by the popularity of the war as the context for popular contemporary writing.[11] Authors such as Pat Barker and Sebastian Faulks have no personal war memories as combatants or non-combatants but they have been energised to write on this theme through the existing canon of war literature. In Ireland, also, there has been a renewed interest in setting creative works within the context of the First World War. Boyce has argued that this trend reflects a response to the onset of violent conflict in Northern Ireland since the late 1960s, and, in part, these works read the war through the lens of current political events.[12] While contemporary writings are important in themselves and offer insights into the significance or insignificance of the war in conjugating present-day identities, in this chapter I will concentrate on works of literature produced by those who directly and indirectly experienced the First World War and who wrote in the shadow of political events taking place on the island in the immediate post-war era.

In the case of Irish literary representation of the war, three observations may be made. First, compared to other combatant states there is a relatively small output of work in the canon of Irish literature. That the Easter Rebellion may have overshadowed the First World War in the literary imagination is significant and will be dealt with in Chapter 6. Second, the historical geography of literary form reveals significantly different approaches to representation between combatant and non-combatant writers. The more experimental and innovative forms often come

---

[10] S. Hynes, 'Personal narratives and commemoration' in J. Winter and E. Sivan, eds., *War and remembrance in the twentieth century* (Cambridge, 1999), 206.

[11] Examples include P. Barker, *Regeneration* (London, 1991); S. Faulks, *Birdsong* (London, 1993); F. McGuinness, *Observe the sons of Ulster marching towards the Somme* (London, 1986).

[12] Boyce, *The sure confusing drum*.

from those not directly fighting on the front. While pastoral themes sometimes pepper the work of soldier-writers, more modernist approaches are found in the writings of non-combatants. Third, the post-war context in which this literature emerged had an impact on its circulation and reception by Irish audiences. While the commemoration of the war through monuments and memorial rituals was informed by the ongoing political dilemmas faced in Ireland, the production and dramatisation of the war through literary commemoration was also affected by this context. Hence this chapter will examine in some detail Sean O'Casey's war play *The Silver Tassie* which is the most important literary work on the war by an Irish writer of the time. The controversy surrounding the production of the play will be analysed because the political context was as central to that discussion as the artistic merits of the play. In terms of the content of the play this chapter will stress the imaginative moving between two spaces – the Home Front and the war front – which anchors the play's depiction of war. The religious tropes that underpin the narrative sequences will be examined and the ways in which a moral geography of war is subverted through a collapsing of the standard bi-polarities of here/there, home/front, sacred/profane will be highlighted. The chapter also engages with the literary works of Irish soldier-writers. The use of more conventional styles of narrative and the significance of personal biography in their scripting of the war experience will be emphasised. Although rarely works of great artistic merit, they nevertheless reveal how the personal biographies of Irish men serving at the Front are woven through common themes. Catholicism, alienation from home and a sense of common cause pervade their work. Their marginal status as writers in the literary canon perhaps mirrors the status of the war in the construction of a national commemorative tradition, especially in the Irish Free State. Collectively these works offer us an opportunity to see how the war got played out and mediated through a literary imagination. In addition, they offer insights into how these writers offered a view of war that, in many ways, reflected the conflict of identity that the war engendered in general and crystallised in particular for Irish Catholic soldiers serving in the British army.

Dramatising war: Sean O'Casey's *The Silver Tassie*

Although the literary mind had conventionally used art to intensify and dramatise the mundane realities of everyday life, the war itself had created such an intense set of human relationships and sufferings that the writer's task was to communicate these heightened and exceptional circumstances in ways familiar to an audience. For soldier-writers, this task was all the more difficult in light of their eyewitness experience of the horrors of the trenches. For Sean O'Casey, however, it was from the Home Fronts of Dublin and London that his war play was conceived and executed. That fact, however, would contribute to a lasting controversy and bitterness over the production of the play, *The Silver Tassie*, in Dublin's Abbey Theatre. The rejection of the play by the directors of the Abbey in 1928 resulted in a vitriolic exchange in the letters columns of the Irish, London and New York press, giving the episode an international flavour. W. B. Yeats (the Abbey's director), who publicly rejected the play, levelled two primary criticisms: O'Casey had not experienced the war directly, and the play lacked a dominant central character. Yeats claimed:

> But you are not interested in the Great War; you never stood on its battlefields or walked its hospitals [and] there is no dominating character, no dominating action, neither psychological unity nor unity of action, and your great power in the past has been the creation of some unique character who dominated all about him and was himself a main impulse in some action that filled the play from beginning to end.[13]

While the war had generated an enormous literary output from combatant soldiers, the overall weakness of Yeats' criticism exposed him to a virulent response from O'Casey:

> Do you really mean that no one should or could write about or speak about a war because one has not stood on the battlefields? Were you serious when you dictated that – really serious, now? Was Shakespeare at Actium or Philippi? Was G. B. Shaw in the boats with the French, or in the forts with the English when St Joan and Dunois made the attack that relieved Orleans? And someone, I think, wrote a poem about Tir na nOg who never took a header into the land of youth. And does war consist only of battlefields?[14]

Trudi Tate in her study of war fiction draws attention to the distinction between witnessing the trauma of war and participating in it, when she claims that many had lived through the war both as soldiers and civilians but had only partially seen it 'through a fog of ignorance, fear, confusion and lies'.[15] Direct participation in war can offer as partial a view as watching from the Home Front. Sean O'Casey had certainly borne witness to the war, if not as a soldier, certainly as a creative spectator. While the acrimony over the production of *The Silver Tassie* would dog the relationship between O'Casey and Yeats, the precise source of the latter's objections remains obscure. That he did not like the play and thought it unsuitable for production at the Abbey was a view that remained un-revised during the course of his life.[16] On the other hand, George Bernard Shaw argued: 'It is literally a hell of a play; but it will clearly force its way on to the stage and Yeats should have submitted to it as a calamity imposed on him by an Act of God . . . Besides he was extraordinarily wrong about the facts.'[17] The controversy over the play would result in the permanent exile of O'Casey from Ireland. Perhaps more importantly, the public airing (initiated by O'Casey) of the literary dirty linen about the play would mean that the critical response to the first and some of the subsequent productions

---

[13] Letter by W. B. Yeats, *The Irish Statesman*, 10, 9 June 1928.
[14] *Ibid.*
[15] Tate, *Modernism, history and the First World War*, 1.
[16] S. Cowasjee, *Sean O'Casey: the man behind the plays* (London, 1963).
[17] A. Gregory, *Lady Gregory's Journals 1916–30*, ed. L. Robinson (London, 1946), 110–11.

of the play would be informed by Yeats' original criticisms. In short, the merits of the play could not be interpreted independently of the Abbey controversy. The stylistic innovation and experimental structure of the play were overshadowed, or at least measured, by the standards of judgement provided by Yeats in his rejection of the play.[18] While the Yeats controversy focused on style, Jeffery reminds us that 'there was an underlying political difficulty in putting sympathetically portrayed British soldiers on the stage of the Abbey Theatre in the late 1920s'.[19] Or at least Irish soldiers fighting in the British army.

In *The Silver Tassie*, O'Casey combines the real and the symbolic, the sacred and the profane, in a characterisation of the First World War which examines the effect of war on the soldiers themselves and the society from which they were drawn. Set in the Home Front (Act I and Act IV), the battle zone of the trenches (Act II) and a transitional space between trench and home in a hospital ward (Act III), O'Casey implicates church and state, combatant and civilian, man and woman in the terror of war. In the Apollo Theatre in London, under the direction of the Canadian, Raymond Massey, and through the support of Shaw and the set painting of Augustus John, the play received its first public performance in October 1929. Although it ran for only twenty-six performances and was thus a commercial flop, it received positive critical acclaim. Many critics continued to echo Yeats' misgivings about the lack of a central unifying character and the expressionistic structure of the second act of the play. The reviewer for the *New Statesman*, for example, commented: '[it] lacks the homogeneity, the essential unity of a really good play',[20] while *The Spectator*'s critic similarly objected to the absence of dramatic unity and the use of expressionism, 'that word, that method, that mistake!'[21] By contrast, the *Irish Times'* reviewer thought the style 'of absorbing interest, and not less interesting because he has not perfected it. Of even greater value is his attempt to break free from the bonds of naturalism by the bold use of verse.'[22] Some recent commentators have reckoned *The Silver Tassie* 'a terrible play. Perversely, but not incompatibly, it is also a masterpiece'[23] or 'arguably the writer's most accomplished play'.[24] The most significant piece of Irish war literature therefore was first performed outside Ireland, viewed by English audiences before it made any impact at home. While Yeats' objection to the play denied it an airing on an Irish stage in the 1920s, perhaps the artistic critique masked a political ambivalence to the content of the play in a post-independence Irish Free State.

---

[18] C. Kleiman, *Sean O'Casey's bridge of vision* (London, 1982).

[19] Jeffery, *Ireland and the Great War*, 95.

[20] J. B.-W., 'The Silver Tassie', *New Statesman*, 3 October 1929, 52.

[21] R. Jennings, 'The Silver Tassie by Sean O'Casey: at the Apollo Theatre', *Spectator*, 143, October 1929, 523.

[22] Quoted in J. Simmons, *Sean O'Casey* (London, 1983), 102.

[23] H. Leonard, 'Aldwych: *The Silver Tassie*', *Plays and Players*, November 1969, 20.

[24] Kiberd, *Inventing Ireland*, 240.

Act I: Confronting war on the Home Front

One of the central unifying characteristics of *The Silver Tassie* is the elaborate use of symbol to study the social and psychological impact of war. O'Casey readily acknowledged this when he claimed:

Yes, The Silver Tassie is concerned with the futile sacrifice of a young Hero in war, and the symbols, the chanted poetry and the ritual of Sacrifice are embedded in the drama... I wanted a war play without Noise...to show it in its main spiritual phases, in its inner impulses and its actual horror of destroying the golden bodies of the young, and of the Church's damned approval in the sardonic hymn to the gun in Act II.[25]

Although he uses symbol and metaphor to 'go into the heart of war',[26] O'Casey's play, in many respects, explores the darkness at the heart of early twentieth-century society. Structured around four acts, the play revolves around the journey of Dubliner Harry Heegan, his family and friends as he moves from working-class Dublin to the Western Front and back again via the hospital wards to his football (soccer) club in Dublin. The opening act, staged in the bedsittingroom of Harry Heegan's home, begins with the anticipation of the return home from the football match of Harry with the winner's cup, the silver tassie. While the heroic setting is being organised by those doing the waiting, O'Casey has Harry's presence dominate the scene even though his triumphant entrance does not take place until half way through Act I. When he finally makes his entrance, he is carried on the shoulders of his team-mates. Although this stage direction did not appear in the original text its introduction by the director Massey in the Apollo production of the play appears to have won the approval of O'Casey and has been used in most productions thereafter. The symbolic significance of the elevation of Harry and the tassie by his girlfriend Jessie 'as a priest would elevate a chalice' establishes the religious idiom that O'Casey would use for the remainder of the play and which would inform the audience's reading of the drama.

The height that Harry had risen to both metaphorically and physically in Act I is juxtaposed to his literal and psychological decline by the end of the play. In the opening act, however, as if to mark his place as hero, Harry instructs his girlfriend Jessie to 'Lift it up, lift it up, Jessie, sign of youth, sign of strength, sign of victory!'[27] The silver tassie in this act may represent Harry's status as champion on the football pitch and the drinking of wine from this cup may mock the communion of Mass, but this ritual will be reenacted in the final part of the play to emphasise the lack of redemption experienced by the soldiers of war. Susie, in Act I, lays down the terms of reference in which the rationale for war is taking

---

[25] Quoted in R. G. Rollins, *Sean O'Casey's Drama: verisimilitude and vision* (Alabama, 1979), 118. Letter to Rollins dated 24 March 1960.

[26] S. O'Casey, *Mirrors in my house: the autobiographies of Sean O'Casey* (New York, 1956), 134.

[27] S. O'Casey, 'The Silver Tassie' in *Seven plays by Sean O'Casey*, selected and introduced by R. Ayling (London, 1985), 194. Although the play was originally published in 1928, throughout the remainder of this chapter this 1985 edition of the play will be used and abbreviated to ST.

place: 'The men that are defending us have leave to bow themselves down in the House of Rimmon, for the men that go with the guns are going with God.'[28] The ritual of drinking the wine of victory in Harry's Dublin home is matched by words which anticipate battle: 'Out with one of them wine-virgins we got in "The Mill in the Field", Barney, and we'll rape her in a last hot moment before we go out to kiss the guns!'[29] The sexual imagery so potent in this scene will contrast with the impotence, spiritual and physical, that will emerge in the context of battle.

The women figuring in this opening act – Jessie the girlfriend, Mrs Heegan (Harry's mother), Teddy Foran's wife, the God-fearing Susie – each play a role in domesticating the scene of the Home Front as though eliding the space between here and there. Unlike O'Casey's women in other plays, critics have pointed to the presence of a 'gallery of predatory women'[30] who surround the action. The portrayal of women in this play mirrors the negative images of women that characterise some of the work of the war poets.[31] O'Casey's use of domestic space as a site of battle between the sexes pre-empts Act II where the action occurs directly in the trenches of the Western Front. To transport us from the domestic front to the battle front, Harry's brightly coloured football gear is replaced with the drab khaki of the soldier's uniform and each of the women assist in the preparation of Harry for battle. The choral chant declares 'You must go back.'[32] O'Casey conveys to the audience that not only are Harry and his team mates participants in war but the entire society is embroiled in the making and maintenance of war. The choral voices dramatise the message and subsume the individual under the collective weight of the group whose moral authority is delivered in the imperative mood. As Harry and his friends step on to the boat which waits to transport them to France, the ship's masthead can be seen as a cross through the window of Harry's bedsittingroom. The image of the cross will dominate the scene set in Act II.

Act II: The ritual of war on the Western Front

The powerful second act of the play prompted the critic Granville-Barker to comment that: '[O'Casey] employs symbolism of scene and character, choric rhythms of speech and movement, the insistence of rhyme, the dignity of ritual, every transcendental means available in his endeavour to give us, seated in our comfortable little theatre, some sense of the chaos of war'.[33] While, for some critics, Act II represents the failure of the play to sustain a coherent structure in terms of plot, character development or action, it is precisely the mingling of realism with expressionism in his articulation of trench warfare which makes this act compelling

---

[28] ST, 196.   [29] ST, 196.   [30] Leonard, 'Aldwych: *The Silver Tassie*', 20.

[31] See, for instance, S. Sassoon's poem 'The glory of women' in I. M. Parsons, ed., *Men who march away: poems of the First World War* (London, 1965).

[32] ST, 197.

[33] H. Granville-Barker, *On poetry in drama* (London, The Romance Lecture, 1937), 25.

and hauntingly disturbing. The physical staging of this act sets the tone for the ensuing dialogue. Set in the trenches somewhere in France, the backdrop is a ruined monastery with a damaged crucifix leaning perilously forward on its pedestal inscribed with the words 'Princeps Pacis' and an image of the Madonna in the stained-glass window of the church. This parodying of the symbolism of the pietà and the ironic use of the image of the prince of peace sets the tone for the characters that occupy the foreground. O'Casey employs an image of a warring landscape that was familiar to the writers and artists of the First World War. Burning chapels and the desecration of religious iconography appear in novels, poetry and paintings of the war and the circulation of these types of images of the landscape of the Western Front would have been reasonably familiar to the audience of the time. Sometimes these images were circulated to denote the barbarism of an enemy, which would attack the very physical foundations of the seats of morality. O'Casey and others also regularly used these images to query the Church's response to war, and its role in its perpetuation and legitimisation.

In the foreground of the set for Act II, a soldier is tied spread-eagled to the wheel of a gun carriage as punishment for stealing poultry. His pose mirrors the crucifix occupying the background. In central position on stage is located a howitzer gun with its barrel pointing towards the enemy along the Front, but literally pointing towards the audience. This piece of military hardware is one of the most enduring symbols of the machinery of the war and its capacity to destroy soldiers' mental and spiritual health as well as their bodies reinforces the increasingly confused distinctions between the human and mechanical vehicles through which war is expedited. The landscape, which O'Casey paints in this act to frame his characters, is bleak and disturbing: 'Here and there heaps of rubbish mark where houses once stood. From some of these, lean, dead hands are protruding. Further on, spiky stumps of trees which were once a small wood.'[34] The soldiers in this act remain anonymous except for Barney who is the only explicitly named character from Act I. All other characters' identities, although somewhat doubling the characters of Act I, are ambiguous.

The most powerful and mysterious character of the scene is the Croucher elevated on a ramp above the other soldiers who hover around the fire. His physical appearance and presence convey a sense of his isolation from those around him but his material and metaphorical presence as a ragged and decaying soldier embodying the Angel of Death or God of War conveys 'both the face and the soul of war'.[35] The Croucher's reversed deployment of biblical phrases delivered through chant continues O'Casey's use of the Mass as an organising framework. While the other characters on stage have lost their individual identity as presented in Act I, the chant of this act is used (as Ellis-Fermor has suggested) as 'a kind of mass subconsciousness'.[36] Although the central character Harry Heegan is never

[34] ST, 200.   [35] Kleiman, *Sean O'Casey's bridge*, 33.
[36] U. Ellis-Fermor, *The frontiers of drama* (London, 1945), 122.

explicitly named in Act II the suggestion that Harry may now inhabit the soul of the Croucher helps to maintain a dramatic tension in this act by uniting the audience in their search for Harry. The reference to items which recall the Home Front (for instance, football matches and the colours of Harry's team in Act I) remind the audience of the connectedness between the soldiers in the trenches and the Home Front. It reinforces O'Casey's theme that the de-individualising effects of war never totally overcome the personal biographies of specific soldiers. As the attack begins, the soldiers converge around the howitzer as 'the only object of veneration, the only help in the hour of death and destruction, the only strong unbroken thing'.[37] The Croucher descends from his elevation to join the other soldiers and in so doing O'Casey has him reveal the 'terrifying incarnation of the God of War' where the Croucher can no longer be distinguished from the other soldiers and we see 'once again the ugly, monstrous, terrifying face of War: the staring empty eyes, the body deformed by the crouching posture, the voice chanting lifelessly in response to their corporal's hymn of praise to the gun'.[38]

O'Casey's transportation of the action of the play from inner-city Dublin to the Western Front involves an intellectual as well as a physical movement. The devices, which are employed to enable the audience to make this journey, are in some ways comparable to the techniques used by the German expressionist playwright Ernst Toller. Although O'Casey denied any conscious attempt to adopt an expressionistic approach[39] the heavy use of ritualised symbol in addition to the combination of abstract and realistic modes of communication resonate with some of the features of expressionist writing. The use of chant to deliver much of the dialogue of Act II, the introduction of a somewhat spirit-like character Croucher, the implication of character doubling from Act I and the persistent and ironic use of religious iconography throughout this act, underline O'Casey's adoption of the abstract as a useful vehicle for rendering the horror of war. While Act II provides us with a glimpse of life on the Western Front, the final two acts of the play transport us back to the Home Front and to the impact of war on combatant and non-combatant alike.

Act III: Recuperating mind and body in a Dublin hospital

The setting for the penultimate Act III in a hospital in Dublin returns the characters and the audience to the Home Front context from where the journey began. However, instead of being staged in the tenement building that the principal characters occupied in the first act, the drama has shifted to a hospital ward adjacent to the

---

[37] A. G. MacDonnel, 'Chronicles, the drama', *London Mercury*, December 1929, xxii.

[38] Kleiman, *Sean O'Casey's bridge*, 37.

[39] In a letter from O'Casey to Rollins the playwright claimed that 'I never consciously adopted "expressionism", which I don't understand and never did,' 24 March 1960. Reproduced in Rollins, *Sean O'Casey's drama*, 118.

hospital garden. The ward is furnished with medical charts, a fireplace, lockers and a statue of the Blessed Virgin which is decorated with the inscription *Mater Misericordiae, ora pro nobis* (Mother of Mercy, pray for us). It is dusk. The anonymity of the characters of the second act is replaced with the individuality of the soldiers in this act 'whose lives have been irrevocably reshaped by the tragedy of war'.[40] Like Act I, the audience is left waiting in anticipation of Harry's entrance, wondering whether he may return as a war hero in a manner reminiscent of his return from the football match in the opening act. The entrance is dramatic and prefaced by Sylvester's unintentionally ironic phrase from the Bible (Samuel 1:19 and 25), 'how are the mighty fallen, and the weapons of war perished'.[41]

In contrast to his triumphant entrance with the silver tassie earlier in the play Harry enters this act 'crouched in a self-propelled invalid chair'.[42] Through the use of this mechanical device Harry engages in a repetitious and purposeless journey around the room: 'Down and up, up and down. Up and down, down and up.'[43] As Tate reminds us, 'Perhaps the most enduring image of the Great War is of the male body in fragments – an image in which war technology and notions of the human body intersect in horrible new ways.'[44] The pain of Harry's predicament and the agony the war has brought to his mental and physical well-being is accentuated by the trite words of consolation uttered during the visitation from his comrades and family. The survival of Barney (at least without sign of physical injury), his award of the Victoria Cross and his liaison with Jessie accentuates Harry's destruction. Susie, now a nurse on the ward, cultivates a distance between herself and her patients by addressing them by their bed numbers rather than their names. This strategy by O'Casey reminds us that the anonymity of the soldier on the front facing an unknown enemy is replicated to some degree when the troops return home to familiar surroundings. It has been suggested that 'The wounded returned soldier became a spectacle in civilian society – a sight of both fascination and dread. He was a paradox: as a soldier, he represented a powerful social ideal of manhood, yet the act of soldiering had damaged the bodily basis of masculinity.'[45]

Bernard Shaw commented that 'the hitting gets harder and harder right through to the end',[46] and Act III brings the viewer close up to Harry's tragic condition. His agitation over the absence of Jessie, and the dwelling on trivialities, which characterise the dialogue for most of this act, builds the audience's anticipation of the pathos of the final act. Although some critics consider Act III too long and extraneous to the central 'plot', Nicoll claims that 'Nothing greater or finer in the modern theatre had been done than the majestically bitter chants at the altar of the gun or the restless, agitated movement of the third act of this play.'[47] The juxtaposition of the religious and the profane is again captured in Harry's bitter statement: 'I'll make my chair a Juggernaut, and wheel it over the neck and spine of

---

40 Kleiman, *Sean O'Casey's bridge*, 41.   41 ST, 218.   42 ST, 218.   43 ST, 219.
44 Tate, *Modernism, history and the First World War*, 78.   45 *Ibid.*, 97
46 Quoted in J. Gassner, *Masters of drama* (New York, 1954), 570.
47 A. Nicoll, *British drama* (London, 1949), 484.

every daffodil that looks at me, and strew them dead to manifest the mercy of God and the justice of man!'[48] As Kleiman suggests, the metaphor of the juggernaut, represented on stage by Harry in his wheelchair, 'is the image of an idol carried by chariot, beneath whose wheels the living are ruthlessly sacrificed'.[49] The image that Harry conveys in breaking the spine of the daffodil as coldly as his own spine was broken, as he denounces God and nature and the possibility of redemption and justice, underscores Harry's isolation on the Home Front. To wish to break each daffodil, to break nature with his mechanised means of transport, brings home the brutality of the industrialised war against humanity and nature from which he has just returned. The mechanised man has become a replica of the soul-less, inhumane machinery of war to which he has been, at once, witness, victim and participant.

Act IV: Dancing to death at the Avondale Football Club

While the Western Front and the hospital have made their mark on Harry Heegan, the final act returns us to the familiar surroundings of home, the Avondale Football Club.[50] In detailed stage directions O'Casey intended a powerful atmosphere for the scene. The centre of the set comprises a room, which is flanked at the rear with an arched entrance into the dance hall. Above the entrance is a scroll reading 'Up the Avondales!' The back wall has a tall, wide window which opens into a garden decorated with shrubs and a sycamore tree. When the scene opens, the curtains into the dance hall are drawn and Simon and Sylvester are outside in the garden smoking and blind Teddy is pacing up and down the path. As the band plays a tune, the curtains are pulled back and the entrance of Barney, holding Jessie's hand, takes place. Barney, in a navy suit, is adorned with his war medals, including his Victoria Cross. Jessie wears a tight-fitting dance gown with a low cut neck. Behind these two enters Harry in his wheelchair, also wearing his medals. All participants are wearing coloured party hats.

The act opens with Harry following Jessie and Barney around the clubhouse. Reminiscent of the opening act, merriment is being made and alcohol consumed. Harry chooses red wine. Even though Harry suggests that even as a 'creeping thing' he is trying to praise the Lord, O'Casey continues the theme of parodying religious devotion where the Great War is concerned. While Barney and Jessie merrily dance – emblematic of vivacity and youthfulness – Harry's inability to use his legs prompts his comment: 'But stretch me on the floor fair on my belly, and I will turn over on my back, then wriggle back again on to my belly; and that's more than a dead, dead man can do!'[51] This image of the serpent, a biblical symbol of temptation and guilt, reminds us of the pathos with which O'Casey

[48] ST, 231.   [49] Kleiman, *Sean O'Casey's bridge*, 43.
[50] The Avondale football club is fictitious but reminiscent of many of the football clubs around the city of Dublin.
[51] ST, 234.

imbues the character of Harry. For some critics, this led to the accusation that Harry was a 'one-dimensional baby from start to finish'.[52] If Harry emerges as a one-dimensional figure, it is the war that precipitates the retardation of his spiritual and intellectual development. The mad marauding around the dance floor in his wheelchair makes Harry appear as 'a figure of nemesis – one of the most familiar guises of War – seeking, in the absence of any other kind of justice, retribution and vengeance'.[53] O'Casey conveys this most potently through the redeployment of the convention of drinking.

The consumption of red wine in this act and Harry's request for the silver tassie to be filled returns us to the true nature of the drinking episode of the first act. This time Harry is conscious of the earlier communion and observes 'red like the blood that was shed for you and for many for the commission of sin'.[54] The symbolic colour of the wine and the use of the term 'commission' rather than 'remission' of sin indicate Harry's awareness of the idolatrous nature of the earlier wine ritual. Although Harry drinks from the cup like Jesus Christ, he feels his emotional separateness both from God and from the people around and can feel neither love nor sympathy towards either. The pathos with which O'Casey represents this scene epitomises the tragedy and the farce, which gives the act its brutal intensity. The juggernaut character of the vehicle which now transports Harry and the frenzied atmosphere it creates is exploited by O'Casey to undergird the audience's witness of the tortured terrain that Harry now occupies and from which he can see no escape. Those around him (characters from the Home Front and seeming survivors of the war) can offer little solace and it is through the character of blind Teddy that Harry slowly comes to reconcile himself with his role in the destruction of life and humanity during the war. In a wickedly ironic exchange the two characters weigh the balance of the injuries meted out to them:

> HARRY I can see, but I cannot dance
> TEDDY I can dance, but I cannot see . . .
> HARRY I never felt the hand that made me helpless.
> TEDDY I never saw the hand that made me blind.
> HARRY Life came and took away the half of life.
> TEDDY Life took from me the half he left with you.[55]

Harry smashes the silver tassie beneath the wheels of the chair suggesting that he is implicated in his own tragic fall. Those who raise their hand to strike others in war may find that the hand invariably lands also on one's own head, a fact hinted at by the chanting chorus of the second act. Having recognised his own culpability in the theatre of violence of the front, Harry achieves a partial re-joining of his heart and soul.

His bitterness at the loss of Jessie to Barney is made more bearable towards the end of the act and, although O'Casey has been accused of creating a gaggle

---

[52] K. Phelan, 'A note on O'Casey', *Commonweal*, L, 7 October 1949.
[53] Kleiman, *Sean O'Casey's bridge*, 44.   [54] ST, 241.   [55] ST, 242.

of predatory women in this play, the choices that Jessie faces are sympathetically portrayed. While Harry may want to crush her with his juggernaut of inhumanity and the spectre of a brutal war, she refuses to be subdued. Neither his accusation of her being a whore or his assault on Barney fully alienate Jessie from Harry and her lamenting cry towards the end of the play – 'Poor Harry!' – suggests not only her pity but also her own sense of loss and the grief of having to choose between the physically and apparently emotionally able Barney and the disabled Harry. Jessie, then, represents the only character in the play who at least makes an effort to blur the boundary of Home Front and battlefront. Although occupying one space, that of Dublin, she does seem partly to appreciate the suffering occasioned by war and her role in that suffering.

As the play closes, O'Casey allows his principal character the possibility of consolation. He is no longer totally isolated from his Dublin family and friends and Teddy seems able to draw Harry back into the world of the living, to home, albeit one where suffering will persist. They move into the garden. For O'Casey, the space of spiritual and emotional reconciliation is a specific type of place – Home – which is set in opposition to the front of Act II or the hospital ward of Act III. Home in this rendition is not so much the tenement building of the first act but the home of nature where the landscape is not destroyed and pockmarked through the machinery of war. The garden, Kleiman suggests, may be more a Gethsemane than an Eden, where redemption may be promised but is yet unrealisable. Harry may have half risen from the depths of despair and O'Casey presents the audience with the possibility of his and our spiritual renewal.

Situating 'The Silver Tassie'

The corporeal body is the site O'Casey chooses to represent the consequences of war. The relationship between the destruction of the physical body and ideas of masculinity has recently been explored and a complex picture emerges with respect to soldiers' attitudes towards their disability, popular responses in the Home Front, state policy and medical discourse.[56] While the crippled body could simultaneously be represented as an icon of bravery and heroism, it could be as easily a focus of pity. For O'Casey, the manner in which he represents Harry's relationship with Jessie, for instance, is at the level of physical desire. He is unlike Sir Clifford Chatterley, in Lawrence's novel, whose wife Connie arrests our sympathy while Clifford's physical paralysis is matched by his emotional paralysis. By contrast, Harry's disability arouses heated passions and his longing for Jessie is physical rather than intellectual or spiritual. His response to his rejection is one of jealousy and a jealousy of the flesh rather than of the soul. Cowasjee observes that 'To lift Harry to unselfish heights would be to make him a martyr... but it would thwart

---

[56] J. Burke, *Dismembering the male: men's bodies, Britain and the Great War* (London, 1996); K. Verdery, *The political lives of dead bodies: reburial and postsocialist change* (New York, 1999).

the dramatist's purpose and lessen the revulsion against war.'[57] The expressionist·
structure and theme of the play has been likened to Ernst Toller's *Hinkemann*,
where the chief characters of both are emasculated by the war, although critics
have suggested that Toller's play renders more transparent the spiritual destruction
of its hero. By contrast, the most famous of English-language war plays, R. C.
Sheriff's *Journey's End*, bears few resemblances to O'Casey's drama. The latter
referred to it as a play of 'false effrontery', but for the public the play was a huge
commercial success.

This rendition of the First World War has aroused widespread and contradictory
analyses. For Dublin audiences of the Abbey's 1935 production, Cowasjee suggests
that it was interpreted as a 'travesty of the Mass ... it was an insult to the Christian
faith and its proudest possessions'.[58] Catholic journals and newspapers represented
this view, one which was never at issue in London reviews. Perhaps this reveals
more about Ireland of the 1930s than about the quality of the play. Opinions are still
divided over the merits of the play. For some, it is 'disastrous in its present form,
[but] is interesting in substance and exciting in many of its local effects'.[59] For
others, 'The symbolic second act is indeed one of the finest things that O'Casey
has written and it is the very soul of the play.'[60] One analyst has commented
that 'it does not deserve to be thought of as a flawed masterpiece, but as a play
which can finally overcome all flaws to achieve that perfection of form which is,
and has always been, its birthright'.[61] The powerful experimental structure of the
play, the controversy which has surrounded its production, the use of the Catholic
Mass as the organising symbolic framework of the play, all contribute to making
it a distinctly Irish dramatisation of the war and a commemoration of it. At the
same time, the play shares the more universal messages that characterised other
representations of the war. As Hynes puts it, 'war annihilates the past selves of
young men, changes them so utterly from youths into soldiers that a return to a
past life is impossible; and then at the end, it dumps them into the strange new
disorder that is peace, to construct new lives'.[62] Those new lives for Irish soldiers
had to be constructed out of radically different political circumstances.

Combating war: Irish soldiers as writers

In comparison to their English counterparts, the output of other Irish writers of the
Great War is fairly slim. There is a tendency towards the autobiographical and, in
general, they are less experimental than O'Casey. The experience of the trenches,
the yearning for home, their religious background and the sheer brutality of the
warfare dominate their work. While many of these writings, especially the novels,
are of limited artistic merit they do offer a flavour of the common themes which
occupied the serving soldier's mind and the narrative style employed resonated

---

[57] Cowasjee, *Sean O'Casey*, 125.   [58] *Ibid.*, 129.   [59] Simmons, *Sean O'Casey*, 115.
[60] Cowasjee, *Sean O'Casey*, 135.   [61] Kleiman, *Sean O'Casey's bridge*, 48.
[62] Hynes, 'Personal narratives', 218.

· more with traditional forms of expression than the modernist influences found in O'Casey. The personal biographies of these writers are important also because they reveal some of the contradictions faced by Irish men serving in a war they increasingly knew would be unpopular at the Home Front.

Irish war poets

If Sean O'Casey represents the established writer with no combatant experience who takes the war as a theme for his work, Ledwidge represents the tragic figure of the young writer and enlisted soldier who fought on the Western and Eastern fronts until his death in 1917. Francis Ledwidge's life encapsulates some of the ironies and contradictions facing Irish men in their decision to volunteer for service, and his memory also bears witness to the complexity of his attitudes towards Ireland's role in the Great War. This is most vividly expressed in popular memory where Ledwidge is best remembered as the poet who wrote 'Thomas MacDonagh', an elegy to one of the executed leaders of the 1916 Rising. The fact that he was a poet who served and died in France and at times wrote about the experience is rather less well documented or acknowledged.

Francis Ledwidge was born on 19 August 1887, the second youngest in a family of three sisters and four brothers. His father died when he was four years old and his mother brought up the family alone doing housework and fieldwork until well into Francis' adulthood. He was born near Slane in County Meath and spent his young adulthood working in the copper mines at Beauparc and as a road-ganger in County Meath. From an early age, Ledwidge was interested in trade union politics and was one of the founding members of the Slane branch of the Meath Labour Union. When the State Insurance Act became law in 1912, the County Labour Unions had to undertake considerably more work and they decided to appoint a paid secretary. Ledwidge got the post for one year in 1913 as a stand-in for the permanent secretary, James P. Kelly.[63] Ledwidge combined an intense interest in labour politics with a close intellectual and personal friendship with the local Unionist peer, Lord Dunsany, whom he met in 1912 when he was 25 years old. Dunsany had a conventional education associated with the aristocracy: public school in England followed by Sandhurst, a commission in the Coldstream Guards and a tour of duty in South Africa during the Boer War. In addition, Dunsany maintained a passionate interest in the arts, writing several books as well as collections of poetry. It was as poets that Dunsany and Ledwidge shared their most common interest. Dunsany facilitated Ledwidge's work by providing him with access to his library, study and financial assistance and actively encouraged his writing talents. He introduced him to the Irish literati of the time (including Oliver St John Gogarty, James Stephens and Thomas MacDonagh), and he assisted in the editing and publication of collections of Ledwidge's poetry.

---

[63] A. Curtayne, *Francis Ledwidge: a life of the poet* (London, 1972). This is the most definitive biographical account of Ledwidge's life and his work.

Coupled with his literary and labour activities, Ledwidge held strong views with respect to the national question and shared the republican aspirations of the leading revolutionary figures of the day. Curtayne comments that Ledwidge 'brilliantly harmonized in his own life the writings and teachings of Connolly and Pearse, seeing no conflict between Christianity and socialist, revolutionary principles'.[64] Indeed, Seamus Heaney speculates that the geography of Ledwidge's birthplace may in fact reflect the complex political and cultural loyalties expressed by the writer. Heaney writes:

[T]he poet, was actually and symbolically placed between two Irelands. Upstream, then and now, were situated pleasant and potent reminders of an Anglicised, assimilated country... The whole scene was as composed and historical as a topographical print, and possessed the tranquil allure of the established order of nineteenth century, post-union Ireland. Downstream, however, there were historical and prehistorical reminders of a different sort which operated as a strong counter-establishment influence in the young Ledwidge's mind... In a fairly obvious way then, the map of the field of Ledwidge's affections reflected the larger map of the conflicting cultural and political energies which were operative in Ireland throughout his lifetime.[65]

Certainly Ledwidge drew heavily on the pastoral landscape of his youth as a source of inspiration for his poetry. Ledwidge's commitment to the nationalist cause surfaced when the call for the establishment of the Irish Volunteers was proposed at the Rotunda meeting of November 1913. He and his brother Joe were among the founders of the Slane corps and he was elected secretary, a duty that brought him to Manchester to raise funds and to establish a branch there. Much of his spare time was now devoted to training and organising the Slane group. In early 1914 his first collection of poems, *Songs of the Fields*, was at proof stage and Lord Dunsany was preparing his introduction to the collection. He was elected to Navan Rural District Council and Board of Guardians and he took his public responsibility seriously, although his future employment remained uncertain. News of the assassinations in Sarajevo provoked little interest in rural Ireland, although the declaration of war by Austria-Hungary on Serbia warned the leaders of Europe that a major confrontation might be imminent. When Britain declared war in August 1914, Lord Dunsany immediately went to the nearest recruiting station in Dublin and with the rank of captain was posted to the Fifth Battalion of the Royal Inniskilling Fusiliers.

As outlined in Chapter 2, Ireland's response to the war was initially ambivalent and much coloured by the local political circumstances. At a meeting of the Slane corps there was resounding support for Redmond's call to service, with only six men, including the two Ledwidges, opposing the resolution.[66] Indeed, nearly all the Meath Volunteer force sided with Redmond. At a meeting of Navan Rural

[64] *Ibid.*, 56.
[65] S. Heaney, 'Introduction' in D. Bolger, *Francis Ledwidge: selected poems* (Dublin, 1992), 12–13.
[66] Curtayne, *Francis Ledwidge*, 77.

Council, which Ledwidge attended, he dissented from a resolution which supported Redmond's actions and in a meeting of Navan Board of Guardians (whose membership was identical to the Rural Council), Ledwidge continued his protest. In a bitter exchange he claimed: 'In the north of Ireland the recruiting sergeants have been saying to the men "Go out and fight with anti-papal France." In the south of Ireland they will say, "Go out and fight for Catholic Belgium." '[67] Other Guardians accused Francis of being pro-German and unpatriotic, perhaps a coward.

These details from Ledwidge's life serve to illustrate the dilemma that many Irish men faced when the call to arms was declared. Despite his open hostility to Ireland participating in the war, on 24 November (five days after his meeting with the Board of Guardians) he enlisted with the Royal Inniskilling Fusiliers at Richmond Barracks in Dublin. His justification was as follows: 'I joined the British Army because she stood between Ireland and an enemy common to our civilisation and I would not have her say that she defended us while we did nothing at home but pass resolutions.'[68]

Ledwidge's literary output was comparatively slight. Many of his poems have no direct reference to the war. In that sense, the pastoral idiom which dominated much of his work drew inspiration from his home base of rural Meath rather than from his experience of the war in Gallipoli, Serbia and the Western Front. According to Seamus Heaney, his first collection, published in the summer of 1914, displayed elements of 'pretentiousness or archaic poeticisms'.[69] It was with his two later collections *Songs of Peace* and *Last Songs* (the latter published after his death in 1917) that Ledwidge's poetry achieves maturity and that he employs some of the Gaelic techniques of assonance and rhyme learned through his interactions with other Irish poets. Even then, few of his poems directly deal with the war, and his status as a war poet, Heaney suggests, emanates from the combination of 'tendermindedness towards the predicaments of others with an ethically unsparing attitude towards the self'.[70] The poem 'My Mother', written while in a hospital bed in Egypt, conveys something of this attachment:

She came unto the hills and saw the change
That brings the swallow and the geese in turns.
But there was not a grief she deemed strange,
For there is that in her which always mourns.[71]

The personal circumstances of his mother's life, combined with a more common theme of suffering, infuse this poem. Ledwidge returned home on recuperation leave in the immediate aftermath of the Easter Rising. In a letter to a Belfast soldier friend, Ledwidge commented that 'I had a hard graft in Suvla and Serbia',[72] one of his few recorded comments about the conditions of war. Upon his return Ledwidge

---

[67] *Ibid.*, 81.   [68] Quoted in *Ibid.*, 83.
[69] Heaney, 'Introduction' in Bolger, *Francis Ledwidge*, 16.
[70] *Ibid.*, 20.   [71] Bolger, *Francis Ledwidge*, 53.   [72] Curtayne, *Francis Ledwidge*, 95.

composed his most famous poem 'Thomas MacDonagh', which critics applauded as 'Ledwidge's first encompassing of profound lyric mastery'.[73]

His maturity of style emerged then from the impact of the Easter Rising in Dublin rather than his direct experience of war on the Eastern Front. Spending time with his family in Slane reminded Ledwidge of earlier romantic encounters and prompted the writing of a number of poems in memory of the death of an early love. Ledwidge was ordered to report for duty in Derry on 18 May but as he had lost days in transit from Manchester he felt entitled to an extension of leave. To secure an extension he had to report to Richmond Barracks where he had enlisted a few years earlier. As a site associated with the rebellion and the place where some of the leaders were sentenced to death, it took on a new meaning for Ledwidge and led to an altercation with a superior officer. Upon arriving late in Derry, Ledwidge was court-martialled and his lance corporal's stripe was removed. Ledwidge's own account of the episode is captured in his poem 'After the Court Martial'. The final stanza is as follows:

> And though men called me a vile name,
> And all my dream companions gone,
> 'Tis I the soldier bears the shame,
> Not I the king of Babylon.[74]

While the compilation for the second book of verse is strongly autobiographical and the titles indicate his itinerary of the war, many poems deal only indirectly with his experience. With his love of nature and pastoral imagery, Ledwidge's poems frequently omitted direct reference to the war. His final collection of poems, published after his death in 1917 in France, do convey some of the poet's attitude towards the conflict and one detects his increasing tiredness of the pain and drudgery of the experience. In 'Soliloquy', he writes in the final stanza:

> It is too late now to retrieve
> A fallen dream, too late to grieve
> A name unmade, but not too late
> To thank the gods for what is great;
> A keen-edged sword, a soldier's heart,
> Is greater than a poet's art.
> And greater than a poet's fame
> A little grave that has no name,
> Whence honour turns away in shame.[75]

The inability of the poet truly to capture the pain of the war underwrites the sentiment of this piece. Whilst simultaneously acknowledging the bravery of the individual soldier but the culpability of those orchestrating war, Ledwidge is replicating many of the sentiments of the war poets.

---

[73] J. Drinkwater's comment quoted in Curtayne, *Francis Ledwidge*, 156.
[74] Bolger, *Francis Ledwidge*, 66.   [75] *Ibid.*, 74.

By December 1916, Ledwidge was again called back to the front in France and he left Folkestone harbour to cross the channel on 26 December. In January, in a letter to the poet and critic Katherine Tynan, he commented: 'I am a unit in the Great War, doing and suffering, admiring great endeavour and condemning great dishonour.'[76] Some of his loneliness as a soldier returned to the front is captured in the poem 'In France' written in February: 'The hills of home are in my mind, And there I wander as I will.'[77] For many soldiers this was a familiar sentiment as they remembered the landscape of home but for Irish soldiers perhaps the ambivalence of their position in the British army made such sentiments all the more important. In a letter to Professor Lewis Nathaniel Chase (University of Wisconsin) the divided allegiances of the Irish soldier are evident: 'I am sorry that party politics should ever divide our town tents but am not without hope that a new Ireland will arise from her ashes...I tell you this in order that you may know what it is to me to be called a British soldier while my own country has no place amongst the nations but the place of Cinderella.'[78] In July of that year Francis Ledwidge was killed.

Other Irishmen who served in the First World War also produced verse about their experience. For instance, Thomas MacGreevy from Tarbert in County Kerry served as an artillery officer during the war and subsequently met and befriended James Joyce and Samuel Beckett in Paris. Working as a Catholic modernist, MacGreevy won the admiration of Beckett who thought his work important since 'it is the act and not the object of perception that matters'.[79] Although he did not produce a huge collection of war poems, 'De Civitate Hominum' conveys the dislocation of the soldier in the trenches, a dislocation replicated in the style employed by MacGreevy:

> I cannot tell which flower he has accepted
> But suddenly there is a tremor,
> A zigzag of lines against the blue
> And he streams down
> Into the white,
> A delicate flame,
>
> A stroke of orange in the morning's dress.
>
> My sergeant says, very low, 'Holy God!
> 'Tis a fearful death.'
>
> Holy God makes no reply
> Yet.[80]

---

[76] Curtayne, *Francis Ledwidge*, 170.
[77] *The complete poems of Francis Ledwidge*, ed. Lord Dunsany (London, 1919), 269.
[78] Quoted in Curtayne *Francis Ledwidge*, 180.
[79] Quoted in Kiberd, *Inventing Ireland*, 461.
[80] T. MacGreevy, *Collected poems*, ed. Thomas Dillon Redshaw (Dublin, 1971), 17.

The theme of soldiers being dislocated, out of place, in the metaphorical sense, and in the literal sense of being located in a foreign land is also found in his poem 'Nocturne' dedicated to 2nd Lieutenant Geoffrey England Taylor:

> I labour in a barren place,
> Alone, self-conscious frightened, blundering;
> Far away, stars wheeling in space,
> About my feet, earth voices whispering.[81]

As a Catholic Irishman soldiering in the trenches of the Somme, MacGreevy advances beyond the popular pastoral Georgianism common to many writers and focuses instead on the psychological and physical alienation precipitated by the war. MacGreevy survives the ordeal and spends subsequent years in Paris, as a critic in London in the 1940s (where he produces a few biographical works),[82] and as director of the National Gallery in Dublin from 1950 onward. His poetic output all but ceases on his return to Ireland and in some respects his potential as an important poet goes unrealised.

Irish war novelists

While drama and poetry were popular media in which to write the experience of war, the novel also became a valuable vehicle in which to convey the war to a wider audience. War novels began to appear six months after the war was declared and were published with greater and greater frequency. Many early novels continued the tradition of the boys' heroic adventure stories in which the Great War was simply another English battle where the virtues of the English outstripped the vices of the enemy. These types of novels served both as recruiting tracts and as instant history books.[83] As the war progressed, however, new themes and devices of narration would emerge and in the post-war years some great literary novels appeared. In Remarque's *All Quiet on the Western Front* (1929) and in Ernest Hemingway's *A Farewell to Arms* (1929), the novel became one of the most influential means of translating the war. The vast number of war novels published in the decade after the conflict made it a genre popular with publishers and audiences alike. The public's appetite for the war novel continued into the 1930s. Much of the output was pretty mediocre by literary standards and Irish war novelists were no exception in this respect. The desire to write of the brutality of the war in a hyper-realist manner rendered much of this literature unsubtle and lacking literary finesse, yet it does throw insights into how Irish men responded to the call to arms. Most of these works focus on the spaces of battle – the micro-world of trench warfare – and the physical and spiritual claustrophobia engendered by these sites of conflict. The landscape described then is not the panoramic vision of the distanced eye

---

[81] *Ibid.*, 15.
[82] See R. Welch, ed., *The Oxford Companion to Irish Literature* (Oxford, 1996).
[83] Hynes, *A war imagined*.

but the worm's-eye vision yielded in the mud of what Fussell describes as a 'troglodyte' war.[84]

Liam O'Flaherty and the theme of bestiality

For Liam O'Flaherty the First World War represented one episode in a very colourful and sometimes painful life. Born on 28 August 1896 in Inishmore, the largest of the Aran Islands off the coast of Galway, O'Flaherty was the second son of a large, poor, Irish-speaking family. His father Michael was a small landholder and married Margaret Ganly, a descendant of a family of Plymouth Brethren who came to the islands from County Antrim to build lighthouses in the early nineteenth century. Whilst O'Flaherty's early life was characterised by poverty, he nevertheless displayed scholarship at the local school and was offered a place as a postulant at the scholasticate of the Holy Ghost Fathers in Rockwell, Co. Tipperary where he would be trained to be a missionary in Africa. After four years of education at Rockwell, O'Flaherty refused to take his soutane, returning home to Inishmore before going back again to a diocesan seminary in Dublin. For the second time, however, O'Flaherty rebelled by refusing to take his soutane and instead began attending lectures in 1914 at University College Dublin, where he had won a scholarship.[85] During this period O'Flaherty had become a member of MacNeill's Volunteers and although interested in republican politics, O'Flaherty learned a different ideology at college. For Sheeran, 'The seminarian destined for the priesthood becomes, for a time, an agnostic. The student of Thomistic Philosophy, which ultimately rests on Thomistic theology, turns to Marxism.'[86] These two themes would re-emerge in his literary output in the following decades.

In 1915, abandoning the republican Volunteers, O'Flaherty joined the Irish Guards using his mother's maiden name and enlisting as Bill Ganly. He was sent to Caterham barracks for basic training and this experience opened a new dimension in his character. Having hitherto extolled the virtues of the intellect, O'Flaherty became acutely aware of the necessity of the body to be a good soldier. He stated, 'I who had until then worshipped the mind to the extent of neglecting the body, now worshipped the body to the neglect of the mind.'[87] The precise reasons why O'Flaherty enlisted are ambiguous. He made a variety of contradictory statements on the subject although his comment that he joined because it was 'What an adventurous youth felt impelled to do, not through idealism, but with the selfish

---

[84] Fussell describes the structure of trench warfare and the landscape as one comprised of holes and ditches where the enemy remains largely invisible although constantly in close proximity, albeit hidden from view. Fussell, *The Great War and modern memory*.

[85] For a fuller discussion of O'Flaherty's life see J. Zneimer, *The literary vision of Liam O'Flaherty* (Syracuse, 1970); J. H. O'Brien, *Liam O'Flaherty* (Lewisburg, 1973); P. F. Sheeran, *The novels of Liam O'Flaherty: a study in romantic realism* (Dublin, 1976).

[86] Sheeran, *The novels of Liam O'Flaherty*, 62.

[87] Quoted in O'Brien, *Liam O'Flaherty*, 18.

desire to take part in a world drama'[88] may be as close to the mark as any other claim.

After training, he was sent to France as a replacement – 'a new patch on an old garment'[89] – and the routine and anonymity of soldiering life contrasted with the individuality he had cultivated hitherto. The trenches proved a more boring, sordid world than anticipated; he noted with some irony that the newspaper reports of battles he participated in were far more exciting than the dull engagement he had experienced in the trenches. In September 1917, at Langemarck, O'Flaherty was seriously wounded in a shell bombardment and spent many months in various hospitals before being released from King George V hospital in Dublin in 1918. His military service was now over and he was discharged in 1918 with the note that he was suffering from *melancholia acuta*.[90] O'Flaherty's fiction and autobiographies are all in some way coloured by his experience in the First World War. The fictional characters of his subsequent work regularly go through what O'Flaherty said of himself: 'You have to go through life with that shell bursting in your head.'[91] Depression and fear of impending doom often characterise his major characters as they negotiate a line between intellectual clarity and insanity. The writing of fiction may in some ways have acted as a form of personal therapy for the writer. In the immediate aftermath of the war he moved to London, working as a porter, labourer and office clerk. At age twenty-one, O'Flaherty became a somewhat bewildered civilian, unsure of his position or future in this world. His attitude to the war remained ambiguous. At times he claimed to have loathed the war, while at others he regarded it as a necessary preparation for greatness as mirrored in the war experiences of Tolstoy and Socrates. After some months working in London, O'Flaherty joined a ship bound for Rio de Janeiro and spent the next few years travelling across the Atlantic between North and South America and Britain and Ireland. Of this desire to travel, O'Flaherty remarked:

> Anywhere. Away from Europe, somewhere that had never been desecrated by crucifixes and churches and schools and shops and all the shoddy armament of civilisation, which produced nothing better than cripples screaming at their fate, wishing that they were whole in order that they might join the whole men who were crippled in thousands on the battlefields.[92]

From the early 1920s onwards Liam O'Flaherty's life as a serious writer began. In only one novel did he directly deal with his war experience. *The Return of the Brute*, published in 1929, is of limited merit as an artistic work, but it does reflect a theme that would dominate much war literature, namely, that war turned men into brutal animals. The belief that war subordinated the mind to the body is found in Robert Graves' verse 'Recalling War': 'Our youth became all-flesh and

---

[88] L. O'Flaherty, *Two years* (London, 1930), 161.
[89] L. O'Flaherty quoted in O'Brien, *Liam O'Flaherty*, 18.
[90] Sheeran, *The novels of Liam O'Flaherty*.
[91] L. O'Flaherty, *Shame the devil* (London, 1934), 83.
[92] O'Flaherty, *Two years*, 59.

waived the mind.'[93] Similarly Paul Baumer, the chief character in *All Quiet on the Western Front*, claims 'We turn into animals when we go up to the line, because that is the only thing that brings us through safely.'[94] The theme of war arousing the bestial instincts of humanity finds expression in a wide range of war literature and for many, 'the degradation that stunned, allured, and destroyed its victims also repelled them'.[95]

In O'Flaherty's *The Return of the Brute* the setting is the micro-world of the trenches. The novel focuses on nine men. The central character is Bill Gunn, a giant of a man with a restricted intellect. No one character represents the reflective mind attempting to handle the horrors of war; each character is painted as plain-speaking and simple minded. O'Flaherty treats the narrative as one horror built upon another. Early in the novel one soldier in the dark comments: 'I just stuck my hands into somebody's rotten guts. God! What a stink!'[96] Bill Gunn gradually reaches the limits of his toleration as food and sleep deprivation, the dirt of the trenches, the harshness of the landscape, the anonymity of the enemy and the regulation-obsessed Corporal Williams begin to bring about his total collapse. On 20 March 1917 the soldiers are ordered over the top where they are gradually wiped out without as much as firing a shot. In the course of the attack, they lose their discipline when they are ordered to dig holes in the frozen earth. A cold, stark landscape mirrors the cold dark war. Chaos and brutality dominate the scene as the men are gradually killed, one from drowning in mud, another from six bullets in the chest. The tension between Gunn and the corporal gradually builds up and as the brutality continues Gunn's mind continues to deteriorate. The narrator tells us that 'His eyes became blurred and he had a curious hallucination that the Corporal was becoming transformed into a hairy animal; a brute which he wanted to kill.'[97] O'Flaherty, however, does not confine his beast-like imagery to Gunn; he remarks, that 'It was a struggle between two brutes, and Gunn was the superior brute.'[98] As the final episode of the novel develops and Gunn's collapse is imminent, 'The interior of his body appeared to be full of monstrous sound, the roaring of flames [yet] in some remote part of his body, very distant and faint, a chorus of birds, of many species, singing in beautiful harmony.'[99] Thus, although O'Flaherty uses the animal imagery to capture the savagery of war, he also employs nature at times to represent a world devoid of human intervention and its horrors. As Gunn's anger explodes he finally attacks and strangles the corporal before charging in a fit of madness towards the enemy to be mown down by machine-gun fire. Private Reilly is the only survivor.

Although this novel lacks much literary sophistication, it does represent some of the dominant themes and imagery that emerged from the war. Apart from the

---

[93] Quoted in A. Bonadeo, *Mark of the beast: death and degradation in the literature of the Great War* (Kentucky, 1989), 9.
[94] *Ibid.*, 6.   [95] *Ibid.*, 38.
[96] L. O'Flaherty, *The return of the brute* (Dublin, 1998), 25. Originally published 1929.
[97] *Ibid.*, 55.   [98] *Ibid.*, 117.   [99] *Ibid.*, 135.

surnames of some of the characters, the novel does not make any direct reference to an Irish context. For O'Flaherty, the horror of modern warfare surpassed national or ethnic difference and the debilitating effects on the human character are painted as universal. The absence of a spiritual or intellectual dimension to the novel, however, renders his characterisation of Gunn as a madman whose motives and actions are devoid of meaning less convincing. As John Zneimer has suggested, 'Gunn is more a collection of nerve ends than a significant mentality.'[100] Consequently, the subtle ways in which the soldier experiences moral decay through the actions of warfare is somewhat missed in this novel, as O'Flaherty fails to construct his characters as reflective as well as active individuals from the beginning. The brutalised landscape of the Western Front and the claustrophobic atmosphere generated by inhabiting the confined spaces of the trenches seem to lead to the soldiers' moral deterioration. Perhaps ultimately O'Flaherty believed the physical to take precedence over the intellectual in art as well as in life. The experience of war itself seems to have debilitated O'Flaherty's literary imagination and the poverty of his war novel is perhaps in itself in indictment of the war.

Patrick MacGill: the navvy soldier

Patrick MacGill also had direct experience of the war but as an Irish migrant in England. Born in Co. Donegal as the eldest son of a poor Catholic farming family, MacGill was sent out to the hiring fair at the age of twelve and most of his wages were remitted to his parents. When fourteen years old, MacGill migrated to Scotland, initially working as a farm labourer for the potato harvest but later working on the railways and in construction. From this social background came MacGill's literary inspiration and his reputation as a 'navvy writer' on the publication of his 1914 autobiographical novel *Children of the Dead End: The Autobiography of a Navvy*.[101] The use of the technique of writing 'the non-fiction novel', according to O'Sullivan, 'changes MacGill from literary oddity to popular professional novelist'.[102] From his beginnings as an itinerant labourer writing about the lot of working-class people, MacGill came to the attention of Canon Dalton who found him a position in the Royal Library at Windsor Castle. For a brief period MacGill worked as a cub reporter for the *Daily Express* but he returned to Donegal before the outbreak of the war where his anti-clerical and socialist views were unpopular amongst the political and ecclesiastical establishment.[103]

---

[100] Zneimer, *The literary vision*, 111.

[101] This novel and all MacGill's subsequent writings were published by the London publisher Herbert Jenkins. Jenkins was a publisher of popular literature and was very successful in nurturing authors and in marketing their work. Included in his list of authors were P. G. Wodehouse, the Yorkshire novelist W. Riley and the Irish poet Francis Ledwidge.

[102] P. O'Sullivan, 'Patrick MacGill: the making of a writer' in S. Hutton and P. Steward, eds., *Ireland's histories: aspects of state, society and ideology* (London, 1991), 213.

[103] See entry for Patrick MacGill in Welch, *The Oxford Companion to Irish Literature*, 336–7.

MacGill's writings, although ostensibly dealing with migrant themes, especially the poor Irish economic migrant to Britain and the hybridity attendant to that location, were regularly received by the British readership as exemplars of 'Irish' writing and in Ireland as the worst exemplars of anti-Irish writing.[104] To some extent, then, his position as a writer concerned with the genuine experiences of the migrant is somewhat lost to both audiences. The category – Irish migrant literature – had no resonance with the reading public or literary establishment of that time. Literary historians and critics address MacGill's work in either regional or class terms. For John Wilson Foster, MacGill falls within the category 'Ulster fiction' while others classify him in terms of British working-class fiction.[105]

In 1915 MacGill married Margaret Gibbons, the niece of Cardinal Gibbons, and a pulp fiction writer. He also decided to join the war effort, enlisting in the London Irish Rifles. He served as a stretcher-bearer in France where he was wounded at Loos. He documented the horrors of trench warfare in three works, *The Amateur Army* (1915), *The Great Push* (1916) and *The Red Horizon* (1916). *The Great Push* is the most accomplished of the three. Written in the first person, this autobiographical novel recounts his experience as a stretcher-bearer with the London Irish at Loos. The narrative centres on his platoon and their preparation for going over the top at Loos in September 1915. MacGill employs the language and idiom of the common soldier to frame his narrative and although many of the characters are English by birth, some of their Irish ancestry is brought to bear in this story. The dialogue between the men, in particular Felan, Gilhooley and MacGill, is interspersed with descriptive references to the life and landscape of the front line. MacGill claims that 'The orchestra of war swelled in an incessant fanfare of dizzy harmony. Floating, stuttering, whistling, screaming and thundering the clamorous voices belched into a rich gamut of passion which shook the grey heavens.'[106] The men regularly allude to the purpose and potential futility of the war in the eyes of the ordinary soldier. MacGill notes that from when he left England until the dawn of 15 September (seven months in France) he had never met a German. The anonymity of the enemy continually struck the men in his platoon, as did their submission of individual identity to a collective one: 'Soldiers always speak of "we"; the

[104] His second novel *The Rat-Pit* (London: Herbert Jenkins, 1915) builds upon the tale of migration of his first novel, except now the central character, Norah Ryan, is female. Prostitution is the only viable alternative to starvation for the mother and child. It is made clear in this novel that Norah Ryan's path to prostitution is the logical outcome of processes which began in Ireland and continue when she migrates. O'Sullivan claims that 'MacGill's terrible perception is that the very same forces that propel young Irish men into navvyhood propel young Irish women into prostitution', 'Patrick MacGill', 215.

[105] See J. Wilson Foster, *Forces and themes in Ulster fiction* (Dublin, 1974); R. Sherry, 'The Irish working class in fiction' in J. Hawthorne, ed., *The British working class novel in the twentieth century* (London, 1984); H. Gustav Klaus, *The socialist novel in Britain* (Brighton, 1982).

[106] P. MacGill, *The great push* (London, 1984 [1916]), 68.

individual is submerged in his regiment. We, soldiers, are part of the Army, the British Army.'[107] Although it was an international conflict, soldiers maintained their sense of national identity in the face of an enemy.

In most of the dialogue MacGill maintains an antipathy towards the state and the church who sent men to fight a battle they did not understand, but he also recognises the value of religious belief in the face of the pure horror of trench warfare. The role of the Catholic chaplain, Father Lane-Fox, in providing spiritual guidance to the London Irish is acknowledged. And in a surreal sequence towards the end of the novel, MacGill employs the imagery of the crucified Christ to make sense of a soldier's body hung across wire. Recourse to religious iconography and to the redemptive hope underlying a crucifixion continues the theme that war is particularised along the lines of individual belief systems, and the powerful role of the image of a crucified Christ resonated especially among Irish Catholic soldiers. Each chapter in the novel is prefaced by anti-war verses and by Cockney humour. The final journey for injured Rifleman 3008 P. MacGill is along what he refers to as the 'Highway of Pain' transporting the dead and the injured from Loos to Victoria Station in London. Set in the space of the war zone, the only real reference to the Home Front comes when the body of the soldier is returned. This novel then, written in an almost journalistic style, combines a concern with the overall futility of war and the particular experience of the mainly working-class recruits in an Irish regiment. Herbert Jenkins, aware of the demand for and the popularity of war novels, was supportive of books dealing with a war theme. In fact, in order to squeeze the market to its limit, a series of 'potboilers' from the Jenkins press appeared until the public's interest in the war began to wane.[108]

The war represented an important interlude in Patrick MacGill's life. Set in the context of the autobiographical novel that examines the fate of the migrant soldier, MacGill weaves the distinctly national with the universal. The experience of the common soldier isolated in an alien and seemingly desolate landscape is domesticated through the Catholic chaplain ministering to his flock overseas. O'Sullivan is right to suggest that 'It is as a migrant writer, with all those contradictions and compromises'[109] that we should interpret MacGill. After the war, MacGill had difficulty resurrecting his literary career and in his later life, suffering from ill-health and neglect, MacGill's literary output waned. He died in Massachusetts on the day that J. F. Kennedy was assassinated. The spaces of the migrant worker and, in his case, the Irish working-class migrant in Britain, are obfuscated in his war novels as the spaces of trench warfare transform all men into migrants as they experienced the alien landscape of war.

---

[107] *Ibid.*, 134.

[108] For a list of these works see O. Dudley Edwards, 'Patrick MacGill and the making of a historical source: with a handlist of his works', *The Innes Review*, 37 (1986), 73–99.

[109] O'Sullivan, 'Patrick MacGill', 218.

Conclusion

Like other soldiers, for Irish men, scripting the war in prose and verse was an attempt to make sense of the ferocity of the experience. While writers like O'Casey employed, to good effect, the narrative techniques associated with modernism, others adopted more orthodox writing styles to convey a sense of the brutality of war. What neither modernist nor more traditional modes of expression did was endorse the war uncritically as a heroic act. For everyone it represented a sacrifice that could not be made so intelligible even by recourse to religious or moral arguments. In part, this reflects the broader tensions that the war represented for all states, but in the Irish case, the vocabulary of sacrifice was also confused by a sense of a divided identity for Irish soldiers in the British army. Travelling from a kind of no man's land at home to the no man's land of war made the act of translation doubly troubling. Both the personal biographies of war writers and the tropes that permeate their work indicate their fragile capacity to make sense of the war for popular audiences.

The experienced and talented O'Casey, who occupied the Home Front for the duration of the war, navigated through the spaces of home and war zones more safely than soldier-writers. His abstract rendition of trench life contrasted with the mundaneness and banality of life at home. Drawing his perspective on war from home rather than from direct experience of the trenches enabled him to convey how war is conjugated as much by the actions, attitudes and principles of those occupying the spaces of peace as by those engaged in the bloody battles of the front. His use of a distinctly Catholic eucharistic motif to structure the plot of the play and the parodying of the life of Christ through his embodiment in the character of Harry Heegan drew heavily on his own experience as an Irish writer in a largely Catholic place. Whilst his lack of direct war experience led some to query his credentials for writing a war play, it may be precisely his non-combatant status that contributed to the power of his drama in situating war as a dialogue between the home and battle zones. Rather than carving up the landscape of war in dualistic ways, O'Casey focuses the audience's attention on the spatiality of war as itself one of interconnectedness between the spaces of soldier and civilian.

By contrast, those who narrated the war directly from their experience as combatants generally focused on the separation of home and war front. For poets, the dichotomy found expression in the idealised, pastoral landscape of home and the chaotic and disrupted landscape of war. English writers expressed a sense of their dislocation from civilian life on their return home. For Irish writers, the political context of the Home Front further disrupted their sense of identity as soldiers. For Irish war novelists, the dualism between home and battle front was even more accentuated with their narratives almost exclusively focused on the intensity of the micro-space of trench warfare. Lacking the finesse of an accomplished writer like O'Casey, the war novels pictured the war through the lens of brutality. The tight human spaces of soldiering life coupled with the tight physical spaces of the trenches featured centrally. The sense of impending destruction animated the plot structure. The brutalising effect of the trenches, which turned men into animals, and turned the idea of the just war into obscene barbarity anchored these stories. The body became the site through which the ferocity of trench life was shaped. Together, these works by Irish writers, although of varying literary merit, indicate how the war became imagined for those geographically and, at times, culturally remote from the spaces of military action. There was, however, a bloody conflict taking place much closer to home. Chapter 6 will detail how the Rising of 1916 entered the popular imagination and memory much more powerfully in spectacle, stone and text. The alternative narrative of Irish nationalism got underwritten by the memory makers in the opening decades following partition and independence, and the strength of that narrative in consolidating a sense of Irish national consciousness made it more compelling than the painful memory of the Great War.

6

Remembering the Easter Rebellion 1916

To remember everything is a form of madness[1]

While this book has so far focused on the commemoration of the First World War in Ireland in the context of a variety of narratives of identity, this chapter directly addresses the efforts made to commemorate one of those competing narratives – the Easter Rebellion of 1916. Although the Rising interrupted the tempo of the First World War in terms of recruitment and subsequent public remembrance of the war, it also generated its own set of commemorative questions in the years after the establishment of the Irish Free State. If the rebellion was staged literally on the streets of the capital city in 1916, the performance of public remembrance took place in a host of different arenas and the interpretation attached to these public acts of ritual changed with time. This chapter will examine a few moments in the trajectory of remembrance to highlight some of the key spatial and iconographic motifs employed to generate a national narrative of commemoration. If sites established to the memory of the First World War, at times, represented points of political and cultural controversy, the Easter rebellion also provoked mixed reactions among pro-Treaty and anti-Treaty factions within Ireland.

On Easter Sunday in 1991, seventy-five years after what Yeats described as the birth of a 'terrible beauty', the Irish state commemorated the 1916 Rising at a ceremony at the General Post Office in Dublin. The commemoration was a modest, low-key affair which sharply contrasted with the celebrations of the fiftieth anniversary in 1966. In the weeks leading up to the anniversary the press and television broadcasts queried the wisdom of commemorating the rebellion for fear that it would provide ideological and political succour to Sinn Féin and the IRA.

This debate centred around the effect that the 1966 commemoration had on the nurturing of a nationalist historiography and on the creation of a national landscape of commemoration. The debate corresponded with a more general argument

---

[1] This quotation is derived from the character of Hugh in Brian Friel's play *Translations* (London, 1981).

taking place in academic circles about the nature and purpose of Irish historical inquiry and the status of historical interpretation itself in an Irish context.[2] Nationalist intellectuals were juxtaposed with critics of Irish nationalism and television footage from the 1966 commemoration was analysed for signs of triumphalism and militarism. Although many in the media sought to use the commemoration as an occasion to criticise the fiftieth anniversary celebrations they too avoided much analysis of the actual rebellion itself. It was a debate centred on the representation of memory rather than on the ideals and motives underlying the rebellion itself. While the critique was carried out mainly by the intelligentsia, popular interpretations of the role of the rebellion in creating a sense of national identity remained largely ignored. A survey carried out by the newspaper, the *Irish Independent*, revealed that 65 per cent of respondents looked on the rebellion with pride while only 14 per cent did not. In addition, 66 per cent of the respondents thought that the rebels of 1916 would be opposed to today's IRA violence.[3]

Thus although the commemoration for the seventy-fifth anniversary stimulated public debate about the value of keeping alive the memory of the Easter Rising, the fact that this debate was driven by key media commentators and political analysts independent of public opinion concealed in important ways popular audiences' ability to discriminate between events in the past and official or political exploitation of those events. Ironically, as the desire to forget the bloody rebellion of 1916 was being articulated, the necessity to remember those who died in the First World War was being promoted simultaneously. However, even in the early years of the new state, efforts to commemorate the rebellion of 1916 met with opposition among the various representatives of Irish nationalism. Rivalries between pro-Treaty and anti-Treaty factions in the wake of the Irish civil war informed some of the controversy surrounding the public remembrance of the rebellion. The fissures that had characterised pre-war Irish politics re-emerged in a post-war nationalist context.

In this chapter I will focus on three moments in the making of public remembrance of 1916. Specifically, I address the symbolic significance of the use of an image of Cuchulain as an iconic representation of the rebellion. Second, I examine the ritual of funerary processions in the creation of sites of collective remembrance in the genealogy of nationalist politics. Finally, I analyse the landscape of memory inaugurated by the celebration of the fiftieth anniversary of the Rising. These moments highlight the differences between the First World War and the Easter Rebellion entering the material and ideological landscape, and the public consciousness of Irish society.

---

[2] For a full discussion of key elements in the revisionist debate see C. Brady, ed., *Interpreting Irish history: the debate on historical revisionism* (Dublin, 1994).

[3] The thrust of this argument is made by D. Kiberd, 'The elephant of revolutionary forgetfulness' in M. Ní Dhonnchadha and T. Dorgan, eds., *Revising the rising* (Derry, 1991), 1–20.

The drama of Easter week 1916

If the First World War recruited an army which produced a literature of war, the Easter rebellion was led and orchestrated by a cadre of literary men and women who in the years leading up to the rebellion conceived of the role of revolutionary violence in literary terms.[4] As Foster has commented of the Rising, 'its rhetoric was poetic',[5] and the manner in which it was executed was also dramatic in form and expression. The fact that the rebellion began on Easter Monday, lasted a week, involved the death of 450 people (of whom 116 were soldiers and 16 policemen) and the injury of another 2,600,[6] and that it was geographically concentrated in the capital city, Dublin, gave the episode an intensity which contrasted with the protracted stalemate of war on the Western Front. The choice of the General Post Office, on Dublin's main thoroughfare, as the headquarters of the rebellion added to the aesthetic drama as its location presented few strategic advantages from a military point of view (Figure 28). The launching of a revolution on the main street of the capital city, where the Proclamation of a Republic on behalf of the Provisional Government was read, centred the affair at the heart of Irish political life. It enabled the inhabitants of the city to bear witness to the revolution before them rather than receiving it remotely as was the case for those serving in the war across Europe. While they may have been horrified, bemused and shocked by the acts of violence surrounding them during Easter week,[7] the decision by the authorities to court-martial and pass the death penalty on 90 arrested rebels (sentences which were commuted for 75) and to execute 15 of them in the city's main gaol invited Dubliners to again be witnesses to acts of extreme violence. The circulation of news of the executions through the national press quickly diffused this information around the island. The imposition of martial law and the exercise of the Defence of the Realm Act all contributed to engaging the wider population in the debate about the Rising. The public intervention of well-known people such as George Bernard Shaw, who claimed that the rebels should be treated as prisoners of war rather than as traitors, elevated the rebellion to the international stage and affected public opinion towards the rebels and their fate. The brevity of the conflict, the intensity of the state's response and the proximity of a general election to these events allowed the rebellion immediately to enter the public's imagination and to take on proportions perhaps greater than the material event itself.

The idea of the staging of the revolution as a public drama is found in the writings of the rebels and among literary minds observing the rebellion. The recollections of Easter Monday by the poet, Austin Clarke, convey something of the theatrical effect it had on observers: 'The historic hour existed with its secret, countless

---

[4] F. X. Martin, *Leaders and men of the Easter Rising: Dublin 1916* (London, 1967).
[5] Foster, *Modern Ireland*, 479.   [6] *Ibid.*, 483.
[7] For an overview of popular representation of the rebellion in the days during and immediately after the rebellion see Lee, *Ireland 1912–1985*, ch. 1; Whelan, *The tree of liberty*.

Figure 28 General Post Office, O'Connell Street, Dublin

memories of the past, in and of itself, so that even the feeling of suspense and of coming disaster seemed to belong to a lesser experience of reality.'[8] The fact that a collection of intellectuals rather than men of great military experience led the rebellion perhaps distinguishes revolution from total war and contrasts it markedly with the Great War. The generation of 1916 comprised men who had benefited from late nineteenth-century educational reforms (many of whom ended up in the commissioned ranks of Kitchener's army). While the war yielded a measure of economic prosperity in Ireland, the desire for political independence among the leaders of the rebellion combined with the aspirations of those with stronger socialist principles who, under the influence of James Connolly's Citizens' Army,

[8] Quoted in Foster, *Modern Ireland*, 482.

desiring radical economic reforms, could not wait for constitutional politics to hold sway. The articulation of the guiding vision of the rebellion is found in the writings of its leaders. These were drawn from a largely middle-class intelligentsia: Patrick Pearse, a school headmaster, Thomas MacDonagh, a university lecturer, Joseph Plunkett the son of Count Plunkett. Influenced by the writings of George Russell, Standish James O'Grady, W. B. Yeats and others of the literary revival in Ireland, the young rebels articulated their cause often in the aesthetic idiom of the 1890s learned when many were adolescents.[9] As Kiberd powerfully argues, 'The Rising, when it came, was therefore seen by many as a foredoomed classical tragedy, whose *dénouement* was both inevitable and unpredictable, prophesied and yet surprising.'[10] The publication of Yeats' play *Cathleen ní Houlihan* (1902) in which a withered old hag would only be restored to youthful beauty when men were willing to die for her, or of Standish O'Grady's *History of Ireland: Heroic Period* (1878–80), which in English translation introduced the heroic feats of figures from Celtic mythology, were both to inspire the rebels in the Rising.

The relationship between art and life, between the aesthetic and the material, between the ideological and the political underpins the enactment, textualisation and commemoration of the rebellion. W. B. Yeats would ask, for instance, 'Did that play of mine send out / Certain men the English shot?'[11] In an analysis of the literary imaginings underpinning the rebellion, Thompson suggests that 'Art helps to create a historical consciousness, but once that consciousness exists, it is history itself that becomes the work of art.'[12] The status of the rebels as distinguished artists may be subject to debate but their commitment to an artistic trope to enact their revolution is not: 'they [rebels] offered their lives to the public as works of art. Seeing themselves as martyrs for beauty, they aestheticized their sacrifice.'[13] The inevitability of defeat for the rebels, the inadequacy of the military planning underpinning the rebellion, and the enactment of the rebellion in the main streets of Dublin make the theatrical analogy more persuasive. Indeed Michael Collins, the pragmatic soldier, commented of the strategy of Padraig Pearse: 'I do not think the Rising week was an appropriate time for the issue of memoranda couched in poetic phrases, nor of actions worked out in a similar fashion. Looking from inside... it had the air of a Greek tragedy about it.'[14]

The theatrical body on which the rebellion was clothed was influenced by the drama staged at the Abbey Theatre (the Irish Literary Theatre) in the years leading up to the Rising.[15] The first rebel to be killed in the Rebellion was an Abbey actor,

---

[9] Hutchinson, *The dynamics of cultural nationalism*; Garvin, *The evolution of Irish nationalist politics*.

[10] Kiberd, *Inventing Ireland*, 200.

[11] W. B. Yeats, 'The man and the echo' in *Collected poems* (London, 1982), 393.

[12] W. I. Thompson, *The imagination of an insurrection: Dublin, Easter 1916: a study of an ideological movement* (Oxford, 1967), 116; see also Martin, *Leaders and men of the Easter Rising: Dublin 1916*.

[13] Kiberd, *Inventing Ireland*, 210.   [14] Quoted in Foster, *Modern Ireland*, 482–3.

[15] Kiberd, *Inventing Ireland*.

Sean Connolly. The call to arms in the Great War was regularly couched in terms of honour and the duty to defend the righteous against a morally corrupt aggressor. When war was declared, the literary elite of Britain continued to diffuse this message in poetry and prose.[16] In Ireland also, a call to arms for the Rising had to be made and the idiom of the theatre provided just the vehicle. The fact that the rebellion was small scale, sharing the intimacy of the playhouse, that its principal actors were well-known, that it was staged in 'civilian territory' rather than along organised trench lines, and that the audience was so close to the action, all contributed to the appropriateness of the theatrical metaphor as the guiding aesthetic of the Rising. The choice of Easter Monday to begin the revolution (catching the authorities by surprise) was infused with symbolic significance because it prophesied the outcome of the rebellion in the syntax of the Christian calendar – suffering, crucifixion and final redemption. This Christian theme also found expression in the First World War, but the leaders of the Irish rebellion were stage-managing their war perhaps more self-consciously than the military leaders of Europe. The deployment of the Christian liturgy to underlie the Rising provided the rebels with an ideological link with previous uprisings, and the Irish historical narrative could be conceived in cyclical rather than in strictly linear terms.

The concept that the past comprises a series of historical cycles has antecedents in ancient Greek thought, although the Italian theorist Giambattista Vico (1668–1744) is most closely associated with the full exposition of the idea of history as cyclical. Vico sought to offer a view of cultural history, which emphasised a perpetually recurring pattern. He claimed: 'Our Science therefore comes to describe at the same time an ideal eternal history traversed in time by the history of every nation in its rise, progress, maturity, decline and fall.'[17] While strictly cyclical versions of this theory necessitate a return to the same point, recurrence does not necessarily imply the return of an individual set of historical conditions but the recurrence of the general pattern under a variety of different circumstances: 'What is recurrent is not the civilisation itself, but the pattern which the histories of otherwise disparate civilisations all exhibit.'[18] In terms of conceptions of time, cyclical views of the past can be seen to be analogous with clock time where there is a return to the same point on the clock's dial or to the same point in the rotation of the earth, whereas recurrent theories employ a seasonal analogy. The past corresponds with nature's seasons – spring, summer, autumn and winter – each season returns but each year is not the same year. Seasonal time then has both elements of cyclicity and linearity where there is a one digit annual change to the calendar. The Christian calendar shares this recurrence with the birth, maturity and death of Jesus Christ, rehearsed annually through liturgical rituals. Easter time therefore marks a cycle along an overall linear view of time.

[16] Fussell, *The Great War and modern memory*; Hynes, *A war imagined*.
[17] Quoted in G. Graham, *The shape of the past* (Cambridge, 1997), 146.   [18] *Ibid.*, 147.

In the arena of nationalist political thinking and practice the imaginary of recurrence also has had a profound influence. The cultivation of tradition, the re-enactment of ritual, the exploitation of symbol and the anchoring of myth have all contributed to the cult of politics associated with nationalism, and for the rebels staging their rebellion in 1916 this notion of the cyclicity of Ireland's history was not lost. Previous rebellions had been staged, matured and failed. The Easter Rising was part of this larger historical process of Ireland's recurrent attempts to be finally resurrected out of British dominion. The themes of religious sacrifice were cast in secular code and the rebels combined Christian imagery with the legitimacy of their political cause. Yet, as Kiberd points out, while revolutionaries elsewhere replaced the religious with the secular (especially Marxist-inspired revolutionaries), in respect of 1916 'the religious was never occluded or buried, but remained visible and audible on the textual surface'.[19] The fusion of the religious, the historical and the mythological all contributed to conceiving and executing the rebellion as a dramatic performance. This trilogy similarly affected the representation of the rebellion: 'The whole event has been remorselessly textualized: for it – more than any of its individual protagonists – became an instantaneous martyr to literature.'[20] But it is not solely in the written text that the rebellion was memorialised. In the public sphere of architecture, spectacle and funeral the religious, mythological and historical were embedded in the landscape of the national imaginary and it was this alternative narrative of Irish remembrance which in part undermined the public significance of world war. I first wish to consider the emergence of a sculptural icon of the rebellion.

Cast in bronze: Cuchulain

Rather than choosing to represent the memory of the rebellion through statues of its leaders, the most influential visual icon of the Rising to emerge was the bronze statue 'The Death of Cuchulain', on public display in the General Post Office in Dublin. The well-known sculptor Oliver Sheppard modelled the statue in 1911–12. The centenary celebrations of the 1798 rebellion had brought Sheppard several commissions for sculpting commemorative statues. Although the centenary took place at the end of the nineteenth century, statues were being unveiled around the country for the next couple of decades to mark the significance of that conflict and to respond to the political circumstances of early twentieth-century Ireland.[21] Oliver Sheppard knew Patrick Pearse through his brother Willie who was a student of Sheppard's at the Dublin Metropolitan School of Art. Sheppard had visited St Enda's, the school where Pearse was the principal. The sculptor was

---

[19] Kiberd, *Inventing Ireland*, 211.   [20] *Ibid.*, 213.

[21] For a fuller discussion of the role of the centenary celebrations of the 1798 rebellion see Johnson, 'Sculpting heroic histories', 78–93; O'Keefe, 'The 1898 efforts to celebrate the United Irishmen: the '98 centennial', 67–91; Leerssen, *Remembrance and imagination.*

also acquainted with W. B. Yeats and this acquaintance, as well as the publication of O'Grady's translation of the Cuchulain saga, stimulated his interest in Irish mythology. In an interview with the *Irish Times*, Sheppard commented that when he first read about the story of the death of Cuchulainn, he was struck by its suitability as a sculptor's theme.[22] Indeed other sculptors of the day had also worked with images of the mythological character[23] and the influence of the story found many adherents in the artistic and literary worlds. It is no surprise, therefore, that the leading sculptor of the day found the saga an important creative stimulus.

Cuchulain is the hero of the Ulster cycle and the central character in the mythological story contained in *Táin Bó Cuailgne* (Cattle Raid of Cooley). His heroic deeds and supernatural powers are celebrated in the narrative as Cuchulain defends Ulster in the face of southern aggressors. The youth, maturity and death of this warrior are detailed in a variety of manuscript sources.[24] In stories such as this, 'stress is on martial prowess and the defeat of demonic opponents'.[25] In the final part of the saga – the death of the hero – a wounded Cuchulain straps himself to a pillar-stone and fights until his death. As he dies, Morrígan (the Phantom Queen) settles on his shoulder, incarnated as a crow. The first popularisation of the saga in English by Standish O'Grady's work[26] was followed by the publication of Aubrey de Vere's *The Foray of Queen Maeve* (1882) and Lady Gregory's *Cuchulain of Muirthemne* (1902). The latter had an introduction by W. B. Yeats who regularly used the mythological figure in his plays and poems. By the early twentieth century then the story of this Celtic warrior had widespread popular appeal and was part of the greater European interest in Celtic antiquities.

For Sheppard's statue a professional Italian model from the Dublin Metropolitan Art School posed for the body and James Sleator, a painter, posed for the head. It took the best part of a year for the clay model to be cast in plaster by the Italian moulder Gilles Orlandi from the National Museum. He worked as Sheppard's technical assistant for this project. The plaster cast was exhibited at the Royal Hibernian Academy in 1914 and it won widespread praise. Indeed the Belfast Museum and Art Gallery expressed an interest in buying the piece but, as Turpin has reminded us, 'its nationalist associations may have made it unacceptable to the prevailing unionist orthodoxy of the city art gallery

---

[22] J. Turpin, 'Cuchulainn lives on', *Circa* (1994), 26–31.

[23] The sculptor John Hughes made two sculptures of Cuchulain in the late 1890s, although neither survives. The mythological figure is also captured in the drawing 'The Flight of Cuchulainn' by Patrick Touhy (Hugh Lane Municipal Gallery, Dublin).

[24] The origins of Cuchulainn are found in *Compert Chon Culainn* (Birth of Cuchulain). His young life and boyhood deeds are narrated in *Táin Bó Cuailgne* (Cattle Raid of Cooley). A variety of other manuscript sources narrate his training as a warrior and his courting of Emer while *Aided Chon Culainn* recounts his death. For references to sources see Welch, *The Oxford companion to Irish literature*.

[25] Welch, *The Oxford companion to Irish literature*, 388.

[26] The first English version of the saga is found in Standish O'Grady, *History of Ireland: the heroic period* and *Cuchulain and his contemporaries*, 2 vols. Published in 1878 and 1880.

management'.[27] It was on exhibition during Dublin Civic Week in 1927, before returning to Sheppard's studio.[28] While the plaster cast was known to have existed and been admired during the 1920s, it was not until the twentieth anniversary of the Rising was approaching that the idea of using the statue as a commemorative monument was mooted. Unlike Sheppard's other work, this statue was not specifically commissioned and whilst its artistic merits were acknowledged by the experts, its role in national commemoration emerged from the desire to find a suitable image to represent the spirit of the rebellion for its twentieth anniversary. According to the historian Loftus, 'Sheppard's Cuchulainn is a very exact representation in visual terms of the role assigned to the Celtic hero in Pearse's private mythology.'[29] The sculpture only came to light because a prosperous solicitor, art collector and acquaintance of Eamon de Valera, John L. Burke, brought it to the president's attention. With de Valera suitably impressed, the plaster was sent to Brussels where it was cast in bronze by the Compagnie des Bronzes, and it was placed in the General Post Office in December 1934 where it was inspected by Sheppard.

The choice of the GPO as the exhibition space for the sculpture reinforced this place as the most important site for commemorating the rebellion (Figure 28) and as Turpin claims, 'The GPO in effect became holy ground and Cuchulainn became a martyr's memorial – a sacred war memorial.'[30] The bronze itself was supported on a Connemara marble base whose bronze plaque was inscribed with the Proclamation of 1916. While the sculpture was sited so that it could be seen from all angles, it has been suggested that Sheppard might have preferred it to be viewed only from the front side (Figure 29). The statue was unveiled on 21 April 1935 as part of an elaborate military display along Dublin's O'Connell Street and a roll of honour of veterans of 1916 was announced (Figure 30). Key government officials attended the ceremony, including de Valera, Sean T. O'Kelly (vice-president) and the relatives of the executed rebels. While illness kept Sheppard away from the unveiling, his daughter, Cathleen, attended and heard de Valera's speech. In it the president stressed the necessity to maintain a memory of the Rising. He stated: 'Everyone who enters this hall henceforth will be reminded of the deed enacted here. A beautiful piece of sculpture, the creation of an Irish genius, symbolising the dauntless courage and abiding constancy of our people, will commemorate it, modestly, but fittingly.'[31]

The connections between a character from Irish mythology, the leaders of the 1916 Rising and the political climate of the 1930s were all seamlessly interwoven in this unveiling ceremony. Not all shades of political opinion were satisfied, however, with the choice of symbol. Some republicans complained that, unlike the 1916 rebels, Cuchulain had not fought a foreign enemy (the mythological Fionn MacCumhaill had fought foreigners). A republican periodical complained: 'There is nothing told of Cuchulainn that would make a representation of his death a

[27] J. Turpin, *Oliver Sheppard 1865–1941* (Dublin, 2000), 139.   [28] *Ibid.*
[29] Quoted in *ibid.*, 136.   [30] *Ibid.*, 141.   [31] *Irish Press*, 22 April 1935.

Figure 29 Oliver Sheppard's Cuchulain bronze

suitable symbol for the struggle and sacrifice of 1916.'[32] Others suggested that an actual historical figure would be preferable to a mythological one to represent Ireland's fight for independence. It might have been difficult to select a single leader of the Rising who could embody the spirit of the conflict, however, without drawing attention to rifts within the revolutionary tradition. Similarly, the fact that

[32] Editorial, *United Ireland Journal*, 20 April 1935.

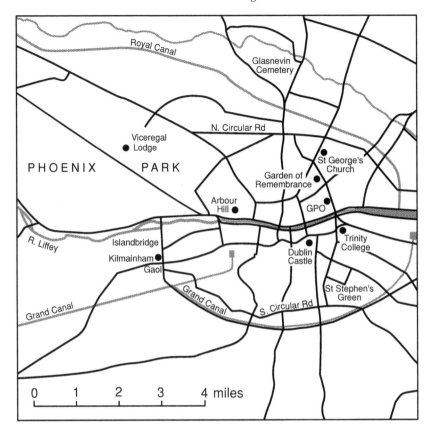

Figure 30 Location map: inner Dublin

Cuchulain defended Ulster rather than the island of Ireland could have had an ironic resonance for those who wished to continue to defend Ulster against any attempts at political reunification.

The potentially contradictory readings of the chosen icon were recognised at the time of its inauguration. The *Irish Times* observed that it was: 'somewhat paradoxical that the warrior who had held so long the gap of Ulster against the southern hordes should now be adopted as the symbol by those whose object it is to bend his native province to their will'.[33] Perhaps for the general populace, however, it was the heroism associated with Cuchulain rather than the geographical details of the Ulster saga that were paramount. His stoical acts of physical resistance, even when out-numbered, perhaps resonated with the public's perception of the Rising. The aesthetic representation of heroic virtue through a statue with a quasi-religious feeling may also have contributed to its popularity. Its similarity

---

[33] *Irish Times*, 22 April 1935.

to images of the Pietá, popularised in the late nineteenth century in Irish churches, made it a recognisable image and the 'fusion of Catholic ideals and revolutionary nationalism – so characteristic of Pearse's rhetoric, was central to its appeal once the 1916 tradition became the dominant ideology of the state as in the 1930s'.[34] The Proclamation beneath the statue offers the political rationale for the sacrifice of 1916. The merging of the mythological and the historical in a single icon of the revolution clouded the distinctions between the two and interestingly simulates the foggy lines between art and life which found expression in the Rising as an act of staged political drama.

The popularity of Oliver Sheppard's statue cannot be underestimated. In the 1937 annual commemoration the Dublin Brigade of the old IRA laid a wreath at the base of the statue, and the image of the memorial has been used in commemorations elsewhere around the country. For both the twenty-fifth and fiftieth anniversary celebrations reproductions of the statue in relief were used in medals and coinage.[35] However, the Cuchulain imagery has been exploited for ventures other than ones of national commemoration and the applicability of the symbol for other purposes underlines the fragility of employing a single icon to do the work of historical interpretation. One commentator has rightly observed, 'the concepts of heroism and achievement, when detached from any religious or political association, could be applied to any new purpose where high attainment and quality were to be proclaimed – as in sport or commercial services'.[36] That Cuchulain can be used to sell jewellery, celebrate the winning of a football match or in Loyalist murals in east Belfast illustrates how rapidly the interpretative apparatus associated with this mythological figure can alter with time and the context in which it is employed. The choice of a mythological figure to embody the spirit of Ireland's most recent revolution mirrors the use of female mythological figures as national icons in other political contexts (Marianne in France or Lady Liberty in the United States).[37] That a male figure rather than a female one, such as Hibernia, was chosen, perhaps in part reflects the masculinity of Pearse's vision of sacrifice. It may also reflect a vision of revolution that was both internal and external (unlike the ones in France and Britain): fought against an external enemy (Britain) but also against internal enemies represented by unionists and constitutional nationalists fighting on the Western Front. Although Sheppard was not commissioned to design a commemorative statue for 1916, his choice of Cuchulain to represent an image of heroism may have hit a chord of familiarity. For de Valera and his associates, who were well aware of the ironies of Irish political memories, they found in Cuchulain an ideal representation of those very complexities, embodied

---

[34] Turpin, *Oliver Sheppard*, 141.

[35] J. Turpin points out that this image was used on veterans' medals distributed in 1941 and that the image also appeared in the commemorative 10 shilling pieces minted to mark the fiftieth anniversary in 1966. Pearse appeared on the other side of the coin, a fact that Turpin notes 'certainly made the point about the close linkage between the two in official State ideology', Turpin, 'Cuchulainn lives on', 29.

[36] *Ibid.*, 30.   [37] Warner, *Monuments and maidens*; Agulhon, *Marianne into battle*.

in the mythological hero. The image of the death of a mythological figure was accompanied by the material burial of historical figures. Let me shift focus now to the funerary rituals associated with the Rising and the development of a national landscape of remembrance.

Burying the dead: the National Graves Association

Burying the dead has played an important role in marking individual sacrifice in war and in maintaining an official memory of the war dead. The tensions between individual and collective memory, religious and secular iconography, state and local acts of burial marked the discussions underpinning the landscapes of remembrance created along the battlefield cemeteries of the Western Front and elsewhere.[38] From the mid-nineteenth century in Ireland, the public burial of important political leaders associated with the independence movement shifted funerary traditions away from individual and local acts of mourning, to displays of 'national' remembrance. The establishment of the first Catholic cemetery in Dublin in 1829 was the culmination of a campaign by Daniel O'Connell to secure the rights of Catholics to control their own funeral services. Up until then, Dublin burials took place in Protestant parish graveyards or cemeteries under Protestant administration. O'Connell used this issue to galvanise support for his Catholic Association and through a series of public debates secured a change in policy. By 1831 the purchase of land in Glasnevin for what was officially called Prospect cemetery had begun (see Figure 30). The cemetery was consecrated in 1832 and the first burial took place soon afterwards.[39] The first large public funeral of an important political figure was the burial of O'Connell himself in 1847. The return of his remains from Genoa, the lying in state of his body in the Pro-Cathedral in Dublin, the Requiem Mass, followed by a large procession to the cemetery at Glasnevin marked this cemetery as the gathering place for the burial of Ireland's nationalist leadership. The *Dublin Evening Post* observed:

For once, Irishmen were unanimous – unanimous in paying a just tribute to the memory of the successful champion of civil and religious liberty. From the highest officer of the State to the humblest citizen, the feeling appeared as if it were a family bereavement that was mourned for.[40]

Following the funeral a permanent memorial was placed in the cemetery to mark the significance of O'Connell. Based on the plans of the antiquarian George Petrie a round tower (based on Early Christian round towers found in Ireland), measuring over 170 feet, was erected and the body of O'Connell was relocated in its crypt at

---

[38] Winter, *Sites of memory, sites of mourning*.

[39] W. J. F. Fitzpatrick, *History of some Dublin Catholic cemeteries* (Dublin, 1900); J. Barry, *A short history of the famous Catholic necropolis* (Dublin, 1932).

[40] *Dublin Evening Post*, 7 August 1847.

Figure 31 O'Connell monument, Glasnevin cemetery

its foot in 1869 (Figure 31). A crowd, estimated to have exceeded 50,000 people, witnessed the event. In his analysis of the O'Connell memorial, the historian Pauric Travers has suggested that there was 'a continuous battle between the Catholic Church and radical political interests for the right to exploit the memory of the dead, [and] the church won this opening battle hands down'.[41] Subsequent funerals would act out this tension between political and Church interests in orchestrating and controlling the rituals associated with them. Parnell's funeral procession (although a Protestant, he was buried in Glasnevin, the cemetery had made provision previously for the burial of non-Catholics) was marked by the absence of Catholic clergy. The crowds that gathered along the route of the cortège, however, indicate that it was the largest public funeral since that of O'Connell

[41] Travers,'Our Fenian Dead', 57.

and that the constitutional politics practised by Parnell won widespread support. Routes of funerals to the cemetery in the north side of the city were varied but tended to straddle the principal streets of the north and south inner city before proceeding to the cemetery. By the early twentieth century, the Irish Republican Brotherhood was re-establishing itself as a force in Irish politics. In 1915, when O'Donovan Rossa was buried in Glasnevin, the physical power of nationalism was reinvigorated, with Pearse offering the graveside oration at the funeral and subsequently becoming a member of the Supreme Council of the IRB. The most cited element of his speech underscores his commitment to blood sacrifice:

Life springs from death; and from the graves of patriot men and women spring living nations... they have left us our Fenian dead and while Ireland holds these graves, Ireland unfree shall never be at peace.[42]

While the ritual of remembrance at Glasnevin was emerging in the century preceding the rebellion in 1916, the particular circumstances of the Rising prevented nationalists from using the cemetery to bury the leaders of the Rising. With the city under martial law all public funeral processions were outlawed and in the years of guerrilla warfare which followed the rebellion, the use of public funerals to commemorate the republican cause was rare. The bodies of the seven signatories of the proclamation of a Republic who were executed at Kilmainham gaol were buried in lime at the military barracks in Arbour Hill (see Figure 30). But the precedent had been established and the placing of Glasnevin on the map of commemorative space as the key site for the burial of nationalist leaders gave the cemetery a key role in the veneration of a nationalist politics. In the years following independence, the cemetery at Glasnevin formed a central node in commemorations of the Rising. On the ninth anniversary of the rebellion, for instance, republican supporters celebrated the occasion by a procession to the cemetery where prayers were recited in Irish at the 'Republican Plot'. Members of parliament attended the ceremony, although no shots were fired over the graves.[43]

The National Graves Association, which grew from the National Graves Committee established in 1926, set about marking graves, maintaining individual plots, and erecting appropriate memorials to those who died for the 'national' cause. The original committee was composed largely of veterans of the Rising and the subsequent war of independence. One of the first tasks of the committee was the establishment of a memorial to the sixteen men, killed in 1916, who were buried in St Paul's section of Glasnevin cemetery. They were buried in a single large unmarked grave and the committee decided that they ought to erect an appropriate memorial. The memorial consisted of a headstone with the names and ranks of the sixteen men engraved on it, and an iron railing about 120 feet in length surrounding the plot. The inscription on the memorial read as follows: 'To perpetuate the memory of members of the Irish Volunteers and Irish Citizen Army who fell

---

[42] Mac Aonghusa and Ó Réagáin, *The best of Pearse*, 134.   [43] *Irish Times*, 13 April 1925.

fighting for the freedom of Ireland, Easter 1916, and whose remains are interred in this Plot.'[44] The memorial was unveiled on Easter Sunday 1929 by Frank Ryan, leader of the Dublin Brigade of the Irish Republican Army. The unveiling took place after the commemorative procession from the General Post Office to the cemetery in Glasnevin.[45] The remit of the National Graves Association extended beyond those killed in 1916 and buried in Glasnevin, and included all deemed to have been 'patriotic' defenders of Irish independence. Thus the Association renovated and cleaned the grave of Wolfe Tone in Bodenstown in Co. Kildare and set about compiling an inventory of burial sites across the country. Although the Association focused particular attention on those killed in the Rising and in the subsequent war of independence, it also mapped the sites of burial of all those over the centuries deemed to have died for the nationalist cause in Ireland.

In 1931 the NGA tried to secure the rights to the 'Republican Plot' in Glasnevin where O'Donovan Rossa and others were buried. As a voluntary organisation, not under the direct auspices of the state, however, the rights over the plot were invested by the government in a separate Plot Committee. They refused to re-linquish their right to decide who could be buried at the site. While the NGA regularly maintained the appearance of the Republican Plot, its ability to control the development of the plot has been limited. Thus although they wished to be the arbiters in adjudicating the worthiness of particular individuals to be laid to rest at the plot, in effect their role was confined to the upkeep of the graves and the use of the site for commemorative activities at Easter time and at other anniversaries. Although the National Graves Association wished to be the principal director in the maintenance of a national memory, it had to compete with the wishes of the state in official commemoration and with other interested parties. In 1932 it published a guidebook to the 'national graves and shrines in Dublin and district'.[46] The book contained the location of graves of those killed in action between 1916 and 1923. It also included a map of Glasnevin cemetery identifying each gravesite of national significance (Figure 32). The foreword to this publication gives a flavour of the political colour of the Association in the 1930s. Written by the chief of staff of the IRA, it states:

National Graves Association deserves praise and congratulations for its effort in making available this permanent record of Patriot graves in and around Dublin, and of the places where many met their deaths in the struggle for national liberty . . .

The day of national commemoration – Easter Sunday – affords annually an appropriate occasion for this patriotic duty . . .

The sacrifices of our patriot dead will inspire our people to safeguard, when it is achieved, the national liberty, for which they so freely gave their lives.[47]

[44] Inscription cited in National Graves Association, *The Last Post*, 2nd edn (Dublin, 1976).
[45] *Ibid.*, 14–15.   [46] *Ibid.*, 17.
[47] 'Foreword' to: National Graves Association, *The Last Post: Glasnevin cemetery being a record of Ireland's heroic dead in Dublin city and county. Also places of historic interest*, compiled and ed. Mary Donnelly (Dublin, 1932).

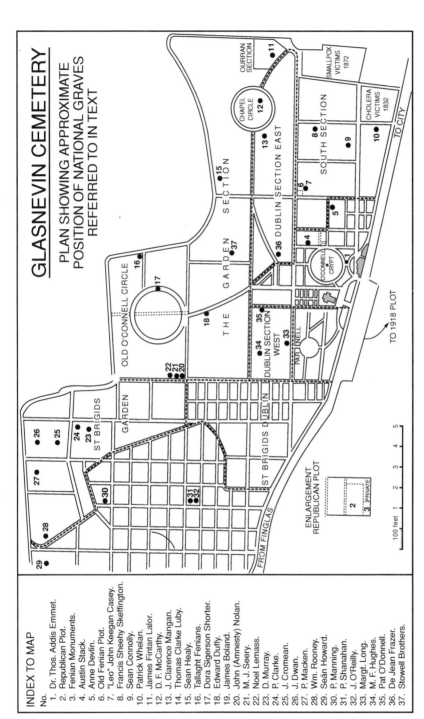

GLASNEVIN CEMETERY

PLAN SHOWING APPROXIMATE POSITION OF NATIONAL GRAVES REFERRED TO IN TEXT

INDEX TO MAP

No.
1. Dr. Thos. Addis Emmet.
2. Republican Plot.
3. Fenian Monuments.
4. Austin Stack.
5. Anne Devlin.
6. Old Fenian Plot.
7. "Leo" John Keegan Casey.
8. Francis Sheehy Skeffington.
9. Sean Connolly.
10. Patrick Whelan.
11. James Fintan Lalor.
12. D. F. McCarthy.
13. J. Clarence Mangan.
14. Thomas Clarke Luby.
15. Sean Healy.
16. Tallaght Fenians.
17. Dora Sigerson Shorter.
18. Edward Duffy.
19. James Boland.
20. John (Amnesty) Nolan.
21. M. J. Seery.
22. Noel Lemass.
23. D. Murray.
24. P. Clarke.
25. J. Cromean.
26. J. Dwan.
27. P. Macken.
28. Wm. Rooney.
29. Seán Howard.
30. P. Manning.
31. P. Shanahan.
32. J. O'Reilly.
33. Margt. Long.
34. M. F. Hughes.
35. Pat O'Donnell.
36. De Jean Frazer.
37. Stowell Brothers.

Figure 32 National Graves Association map of Glasnevin cemetery

A clear link between the commemoration of the dead and the political struggle for reunification of the island is here expressed. While the booklet acknowledges that it is not a complete guide to all graves in Dublin, it does pay particular attention to those interred in the 'Republican Plot' and the old 'Fenian Plot' in the cemetery and those executed and buried in various gaols. It also indicates the location of monuments to Fenian dead. Unlike the Imperial War Graves Commission which worked exclusively to officially bury the dead of the Great War, in Ireland control over the burial, design and commemorative activities permitted at grave sites across the country followed the political and ideological battles that preceded independence between revolutionaries and constitutionalists, church and state, local and national interests. The National Graves Association was one important body in this negotiation of the rights to funerary and commemorative activities. As a voluntary organisation, with strong republican links, the Association regularly ran fund-raising events. The programme for a concert at the Mansion House in late November 1933 indicates that the evening would be made up of a harpist playing a lament to Robert Emmet, the singing of 'My Dark Rosaleen' and the performance of the one-act play *The Gaol Gate* by Lady Gregory.[48] Similarly, in 1941, at the Manchester Martyrs Commemoration Concert, the programme included songs such as 'The Bold Fenian Men' and 'The Battle Hymn' and the play *The Manchester Martyrs: A Propaganda Play in 4 Scenes by Eamonn de Barra*.[49]

In addition to the annual Easter Sunday commemorations, the NGA periodically unveiled monuments they had commissioned and funded. In November 1933, for instance, the MacManus–O'Mahony memorial was unveiled at Glasnevin. There had been an elaborate burial ritual of the Young Irelander, MacManus, in 1861 when his body was exhumed from its grave in the United States and returned to Ireland. This funeral had marked a split between radical revolutionaries of the 1860s and Cardinal Cullen, the Catholic Primate of Ireland. When the body arrived in Dublin, Cardinal Cullen ordered that no Catholic church or clergy should permit the body to lie in state in a church. The cardinal was suspicious of Fenian involvement in the organisation of the event and sought to distance the church from any call to armed revolt. Consequently, the body lay in state for one week at the Mechanics' Institute in Lower Abbey Street in Dublin. The body was then transported to the cemetery in a public ritual where, 'The route to Glasnevin coincided partly with that of O'Connell's cortege and was primarily designed to evoke memories of dead patriots.'[50] Similarly, over seventy years later the NGA invited the public to commemorate MacManus–O'Mahony through the unveiling of the memorial:

[48] National Graves Association, *Souvenir Programme*, Mansion House, 23 November 1937.
[49] National Graves Association, *Manchester Martyrs Commemoration Concert Programme*, Mansion House, 21 November 1941.
[50] Travers, 'Our fenian dead', 58.

A national demonstration to unveil the nation's tribute to the memory of three generations of physical resistance to the conquerors of this country ... The NGA deem this a fitting occasion to invite Irish men and Irish women to participate and thereby demonstrate their allegiance to the ideals for which these successive movements strove so gallantly.[51]

Dr Andy Cooney carried out the unveiling ceremony and the inscription on the memorial included the names of men from different generations. It stated: 'All outlaws and felons according to English law, but true soldiers of Irish liberty; representatives of successive movements for Irish Independence in different ages ... and [they] afford perpetual proof that in the Irish heart faith in Irish Nationality is indestructible.'[52]

Thus, although the NGA concentrated its efforts on those involved in the 1916 Rising and the subsequent war, it also incorporated within their activities those men killed in previous generations. In 1936 a memorial stone was unveiled in Glasnevin to three killed in 1916, and similarly on Easter Sunday 1937 plaques were unveiled at a spot in Dublin where two others were killed in the Rising. The NGA secured the renaming of Moore Lane to O'Rahilly Parade in memory of the dead O'Rahilly. The NGA lobbied the government and local ratepayers for street names in Dublin to be renamed in the memory of Irish men. For instance, Kingsbridge was renamed Sean Heuston Bridge and Stafford Street was renamed Wolfe Tone Street.[53] Thus, in addition to the unveiling activities of plaques and memorials, the NGA also sought to have what it considered to be symbols of British identity removed from the street names and insignia on buildings throughout the city. Although the Dublin Corporation Act 1890 provided the local authority with the right to alter the names of Dublin's city streets, it was not until after independence that the authority would effect many name changes.[54] The NGA did not confine their commemorative rituals to Dublin. They also organised the erection and unveiling of memorials in other parts of the country. In 1946, for example, a memorial was unveiled at Ballinaltin Road, Tramore in Co. Waterford where an ambush had taken place and twelve were killed. The East Waterford Branch of the NGA was founded in 1943 and set about commemorating the twelve men killed in the Old East Waterford Brigade Area between 1916 and 1923 and taking responsibility for the graves of dead republicans from this area. To mark the unveiling, the NGA published a souvenir pamphlet.[55]

The burying of the war dead in Ireland in the aftermath of 1916 differed in several important ways from the laying to rest of the victims of the Great War. First, there

---

[51] Letter from the National Graves Association, *Unveiling of the MacManus–O'Mahony monument*, 4 November 1933. Reprinted in *The Last Post* (Dublin, 1976).

[52] *Ibid.*, 18–19.   [53] *Ibid.*, 23.

[54] Y. Whelan, 'Sackville Street/O'Connell Street: turning space into place, the power of street nomenclature', *Baile* (Dublin, 1997), 4–10.

[55] National Graves Association, *Waterford remembers* (Waterford, 1946). Written by N. de Fuiteóil.

was no overarching state agency vested with the responsibility of erecting war cemeteries equivalent to those on the Western Front. Those killed in 1916 were buried in an *ad hoc* manner; some in plots reserved for republicans in the island's various cemeteries. The vast majority of casualties of the First World War were buried where they fell, inscribing on the landscape of continental Europe Britain's role in the defeat of Germany and in the maintenance of geopolitical stability. The burial of soldiers on foreign soil reinforced a sense of national sacrifice made in the name of a 'just' international war, and the creation of a landscape of remembrance that was literally mapped onto the landscape of battle heightened the sense of Britain's important role in global international relations.

Second, even though most casualties of 1916 were buried in Catholic cemeteries, the Church found itself in an ambiguous position where the celebration of revolutionary violence was concerned. Although there was debate in Britain and other countries regarding the iconographic design of war cemeteries and memorials, the Churches in general were supportive of the war effort: thus the language and symbolism of Christian burial rites did not present a theological dilemma for the Church. In Ireland, however, the staging of a revolution and the practice of guerrilla warfare rendered the Church's position difficult. The Church offered different responses in different contexts depending upon who was controlling the commemorations and the speeches and icons associated with them.

Third, although memorials to the First World War were loosely connected with previous war commemoration, the sheer scale of the event and the huge impact it had on ordinary families across the state in a sense transformed it into a mass event, an occasion for individual and collective mourning in which the state and the civil society were engaged as a unique event in an exceptional set of circumstances. In subsequent generations memorials to the dead of the Great War would be used to commemorate British and Commonwealth soldiers killed in the Second World War and later conflicts. In the case of the 1916 Rising, however, it was explicitly linked to earlier attempts at revolutionary action and to future political ideals. In this sense, the public landscape of remembrance for the casualties of 1916 did not represent the individual mourning of bereaved families but was more clearly connected with the evocation of a national identity. Diverse political lobbies had an interest in the manner in which remembrance would be cultivated and expressed. The role of the National Graves Association in particular exemplifies how the burial of the dead was used to make wider direct links with late eighteenth and nineteenth-century political movements and to continue the republican tradition. The unfinished business of partition weighed heavily in the representation of the role of past revolutionaries and the responsibilities of present-day republicans. The gradual evolution of Glasnevin cemetery as the cornerstone of remembrance cultivated initially in the nineteenth century by the staging of mass funeral processions for O'Connell and Parnell, became, in the twentieth century, more closely connected to unconstitutional nationalism and revolutionary violence. The increase in the number of processions through Dublin's city centre to the cemetery offered a seeming historical continuity between party politicians of the nineteenth century and rebels of the twentieth century. The evolution of a specific route of procession offered the viewing public a sense of historical continuity, where the pavements of the city of Dublin became a map of the revolutionary past of the 'nation'. Unlike the First World War where the geography of death and the burial of soldiers were distanced from home, the Home Front for the rebels of 1916 was the battlefront, and the city of Dublin represented the nexus of that conflict. Reading the spaces of death challenges any suggestion that the universality of death is matched by a universal trope of mourning. Mourning rituals are variously domesticated in terms of orchestration and they are read differently by their viewing publics.

Official commemoration 1966

Although attempts to inaugurate a commemoration of the rebellion in the 1920s, under Cosgrave's government, met with some success, rivalries between pro-Treaty and anti-Treaty factions, and the damage of the civil war created tensions in using the rebellion as a 'credible focus for reconciliation between supporters and opponents of the Treaty'.[56] The first official commemoration at Arbour Hill in 1924 (where the leaders were executed and buried) caused rifts between the government, the relatives of the deceased and virulent republicans. This last group, from the 1920s onward, regularly conducted their own independent remembrance ceremonies at Glasnevin cemetery. The accession to power of Fianna Fail in 1932, under the leadership of Eamon De Valera, offered a greater focus for a unified sense of national remembrance. Nevertheless, some republican factions continued to oppose official commemoration and when de Valera unveiled the Cuchulain statue in 1935, there were processions of rival republicans during the event.

However, the annual commemoration of the Rising of 1916, centred on the General Post Office in Dublin's O'Connell Street, reached its apogee in 1966, the fiftieth anniversary of the Rising. In addition to the large military parade in the city centre, which was witnessed by politicians, veterans, the National Graves Association and the general public, there were a variety of other supplementary events. These included the unveiling of the Garden of Remembrance, the Commemorative Exhibition of Works of Art related to the Rising held at the National Gallery of Ireland, and the broadcasting on Irish television of Hugh Leonard's serial *Insurrection*. With the benefit of hindsight, the celebrations of 1966 have been characterised as triumphalist, militaristic and uncritical. But Kiberd suggests that 'Politicians and propagandists produced a sanitised, heroic image of Patrick Pearse, at least partly to downplay the socialism of Connolly, then attracting the allegiance of the liberal young.'[57] Although the 1960s witnessed a shift in political

---

[56] D. Fitzpatrick, 'Commemoration in the Irish Free State: a chronicle of embarrassment' in I. McBride, ed., *History and memory in modern Ireland* (Cambridge: 2001), 195.

[57] Kiberd, *Inventing Ireland*, 5.

culture towards a more radical, civil rights agenda, especially in the United States, Britain and France, in Ireland they ushered in moves towards modernity under the tutorship of the then taoiseach, Sean Lemass. For some the Golden Jubilee of the Rising presented an occasion in which 'they represented a last over-the-top purgation of a debt to the past, which most of the celebrants suspected would go unpaid'.[58] While the celebrations themselves were on a massive scale, voices of dissent were being raised by cultural and political commentators who recognised the oversimplification of history that the commemorations were heralding. Conor Cruise O'Brien warned of the dangers of triumphalist celebrations when the two most significant national objectives of the new state – political reunification of the island and the restoration of the Irish language – had been quietly shelved.[59] It is unsurprising nevertheless that the state and the population invested so much in this anniversary, given the comparative newness of the state and the fact that the Rising still existed in the living memory of many people around the country. Easter Sunday 1966 provided the opportunity for those in the Irish Republic to pay homage to those who were being presented as the founding fathers of the new state. That those killed in the Battle of the Somme in 1916 would have far fewer public acts of remembrance underlines the relative significance attached to each event by both the state and the population at large.

In terms of public architecture, the most significant monument to the Rising is the Garden of Remembrance in Parnell Square opened by President de Valera, signatory of the Proclamation of the Irish Republic. In 1935, the Dublin Brigade Council of the old IRA suggested to the government that a public memorial be erected in the city centre. The site chosen resonated with historical and political significance. Located in the Rotunda gardens at the northern end of O'Connell Street, the site was where the Irish Volunteers were founded in 1913, some of whom ironically later chose to follow Redmond's advice to join the war effort, while others joined forces to stage the rebellion (see Figure 30). This was also the location where prisoners of the Rising were stockaded overnight on the Saturday evening of Easter week. In addition, this space was in close proximity to the General Post Office (headquarters of the Rising) and within a short distance of the two most significant constitutional nationalists memorialised on O'Connell Street – Daniel O'Connell and Charles Stewart Parnell.

The Garden of Remembrance thus symbolised the road of historical continuity between the early nineteenth century of O'Connell at the southern end of the city's main thoroughfare to the revolution at its northern end. The fact that, mid-way along the street, the 121 foot Doric column of Nelson's pillar, erected in 1808, was blown up in March 1966 perhaps attests to the disjuncture this monument might have represented to some on the eve of the jubilee celebrations. The geography of remembrance could not be disrupted with inconvenient historical reference points.

---

[58] *Ibid.*, 6.
[59] For a general overview of O'Brien's political thesis during these years see C. Cruise O'Brien, *States of Ireland* (London, 1972).

Daithi Hanly's design was chosen from an open competition in 1946 and work on the project began in the early 1960s. The garden is designed in cruciform shape with sunken walkways, pools in the centre and a twelve-foot marble wall enclosing the rear of the garden. At the base of the pool is a green-blue mosaic representing water and carved broken spears similar to those used in ancient battles (300 BC–300 AD). The inspiration for this design was drawn from mythological evidence where warriors would throw their weapons into lakes and rivers after battle. Along the railings of the garden are carvings based on objects held at the National Museum (for instance, Brian Boru's harp). The most significant feature of the garden is the monument designed by the sculptor Oisín Kelly. Drawn from the mythological tale of the Children of Lír,[60] the earlier connections drawn between mythology, history and representation in both the staging of the rebellion and in the Cuchulain statue reinforce the significance of the mythological motif in symbolising the past. Inspired also by Yeats' poem 'Easter 1916' the monument is of four swans from which four human figures arise (Figure 33). This is intended to reflect Yeats' line that men at specific historical moments are 'transformed utterly'.[61] The bronze sculpture was cast in the Marinelli Foundry in Florence. It weighs over 8 tons and is over 25 feet tall. The dramatic and hopeful pose of the figures and the swans in this representation of the rebellion detracts attention from the violence of the Rising itself and suggests the positive hopes embedded in the sacrifice of the rebels of 1916.

The exhibition of over 170 paintings, drawings and sculptures at the National Gallery to celebrate the fiftieth anniversary indicates how the visual arts were also brought into service in the staging of memory. In the foreword to the exhibition's catalogue, it was stated that 'The exhibition does not aim to record aspects of Irish history other than those battles which took place in the slow 300 years of resurgence before 1916 and which were clear armed manifestations of Ireland's nationality.'[62] The omission of people like Wolfe Tone and Robert Emmet is justified by the desire 'to show the spasmodic outbreaks, which represented the will of the people'.[63] Including paintings of the seventeenth-century Battle of Kinsale, the 1798 rebellion and the 1916 Rising, the exhibition assembled an impressive array of the works of the leading artists of the nineteenth and twentieth

---

[60] The tale of the Children of Lír is found in sixteenth-century manuscript form and it relates the story of Lír, the king of the Tuatha dé Danann, who after his defeat and the death of his wife is offered the foster-child of the new King Bobh. She bears him four children, three boys and one girl. After her death Lír takes on a new wife Aífe who grows jealous of Lír's affections for his children and turns them into swans. Under her spell they are condemned to spend three periods of 300 years as swans before being transformed back into human shape when the spell is lifted. They are then baptised. For a fuller discussion of the tale see D. Ó hÓgáin, *Myths, legend and romance: an encyclopedia of the Irish folk tradition* (London, 1990); see also P. Harbison, H. Potterton and J. Sheehy, *Irish art and architecture from prehistory to the present* (London, 1978).

[61] P. Liddy, *Dublin be proud* (Dublin, 1987).

[62] National Gallery of Ireland Catalogue, *Golden Jubilee of the Easter Rising* (Dublin, 1966), 5.

[63] *Ibid.*

Figure 33 Children of Lír, Garden of Remembrance, Dublin

centuries (including Jack B. Yeats, John Lavery and Louis le Brocquy), and it
provided an evocative if not comprehensive collection of Ireland's war paintings.
The exhibition also included a wide array of portraits of the artists themselves. The
focus on the military representation of Ireland's past and portraits of individual
leaders of armed rebellion complied fittingly with the spirit of the anniversary. It
pointed once again to the continuity of armed resistance in the Irish historical record
and it accorded the men and women of violence a legitimate footing in the archive
of national memory. Moreover, the exhibition reminds us of the significance of
war as an inspirational moment for the artist. While the Great War was the subject

matter for many important British artists, the 1916 Rising also proved a fertile ground for stimulating the imagination of Ireland's most significant painters and sculptors.

Conclusion

The re-imagining of the Irish past through the commemorations dedicated to the 1916 rebellion provides us with the context in which remembering those killed in the First World War took place. Three important points emerge from this discussion. The transformation of public space to honour war dead in Ireland had to compete with a pre-existing landscape where nationalist leaders and rebels were already celebrated. Glasnevin cemetery was an appropriate site for the burial of those killed in the 1916 Rising as its nationalist credentials had already been established by the burials of O'Connell, Parnell and others in the nineteenth century. Similarly, the choice of the GPO as the place where the spectacle of commemoration would be staged added to that site's rhetorical significance. The site of remembrance then was where the battle took place and the soldiers actually fell.

The use of mythological figures to represent the spirit of rebellion contributed to an interpretation of it as a form of heroic drama rather than as a single historical and material moment. It transformed the act of remembering from one of mourning the dead to one of celebrating the heroism and beauty of their acts. Violence could be translated into an aesthetic performance embodied in the figures of mythological characters outside, but intervening periodically in, the material world. Commemoration in this guise is not a public act of reconciliation or grieving over individual or collective loss; it is itself a public re-dramatisation of the legitimacy of blood sacrifices. Its meaning is in its overall public effect (declaring Irish independence) rather than the sum of its individual actions. Relatively few individuals were killed so it did not quite touch communities in the same personal ways as did the First World War. In consequence, its public role took on much more significance than its private one of grieving. That ownership of the memory and the rituals of re-membrance of the rebellion was contested in the 1920s, in particular, reminds us of the fragility of creating common threads of identity in post-independence Ireland. The freshness of the civil war in the popular mind could not be easily erased by public acts of commemoration, although by the fiftieth anniversary in 1966 some of these earlier divisions had been bridged.

Finally, the coexistence of mythological personifications of the Rising with a Christian calendar of remembrance presented the audience with a historical narrative that seemed to travel seamlessly from the ancient Celtic world of mythology through to the material achievement of nationhood. This narrative of national commemoration embodied in the Rising of 1916 did compete with efforts to remember Irish soldiers of the First World War, with Unionist remembrance of the Somme, as well as with militant republicans' attitudes towards the Rising in contemporary political affairs. While 1966 represented the high watermark in terms of state investment in the public performance and popular participation in commemorative spectacle, changing political circumstances, north and south of the border, have affected the role of the Rising in the calendar of national remembrance. This is reflected in the contrasting rituals evident at the seventy-fifth anniversary celebrations in 1991, which was a much more low-key affair. By the final decade of the twentieth century the anniversary presented a moment for critical reflection on the events of 1916 rather than an occasion of mass remembrance and celebration.

Even among nationalists the rebellion of 1916 could not, without some controversy, become the unified national moment for the emergence of the Irish state. Ironically, the difficulties identified earlier in this book in finding a cohesive nationalist voice in a pre-war Irish context re-emerged in the post-war period. Partition, civil war, the conflict in Northern Ireland, and the Republic of Ireland's fragile sense of its own national identity all continued to have an impact on the manner in which public commemoration of a bloody conflict could be expressed.

7

Conclusion

The power of the dead to disturb the living is explored in James Joyce's short story 'The Dead'. Gretta's resurrection of the memory of the long-deceased Michael Furey, entombed in a lonely graveyard in Oughterard, in the western county of Galway, displaces her husband Gabriel's cultural and emotional coordinates. From the comfort of their room in Dublin's fashionable Gresham Hotel, Gabriel is confronted with his wife's past. Gabriel's belief that he must look to continental Europe to enhance his intellectual formation, rather than towards the provincial backwater of a city from which he comes or from the west of Ireland where Irish cultural revivalists locate their centre of intellectual gravity, is completely shaken. The revelation that the memory of the dead can inform so much of Gretta's identity challenges his optimism that the future of his marriage and his life rests upon the absorption of European cultural traditions.[1]

Joyce's deployment of memory and spatial categories – the Irish west and the Gaeltacht, continental Europe, Dublin's opera venues, the central plain of Ireland – to anchor his story of early twentieth-century Dublin life, reinforces the contention that time and space, memory and identity get calibrated in fractured ways and the paths of history, nightmarish as they might be, can never be totally relegated to the tomb. This book has attempted to examine the routes by which European history literally travelled into the Irish imagination through the loss of thousands of Irish soldiers' lives during the First World War. The complex spatial politics of the European state system which precipitated the war were matched by Ireland's complex internal geographical alliances and external relations with the rest of Britain and with Europe. The literal and metaphorical construction of memory in Ireland involved a process of making both the space and the time for Irish casualties of the war to be remembered. And this took place in dialogue with a vibrant, and, at times, violent nationalism which engulfed the immediate post-war period.

While geographers have paid increased attention to the politics of memory, their contribution to the field remains still relatively insignificant. This book has

---

[1] J. Joyce, 'The dead' in *Dubliners* (London, 1988). Originally published in 1914 by Maunsel and Co.

attempted to broaden the scope of geographical inquiry into this arena and particularly to emphasise the spatial dimension to memory work. By deploying the metaphor of stages, I have reinforced the view that memory operates within a time–space matrix by re-collecting events of the past while simultaneously re-staging them through a variety of representational practices. These stages – poster, parade, memorial, text – each script the war in the public consciousness even when parts of that public wish to forget. Roland Barthes has been particularly influential in drawing my attention to the role of spectacle in everyday life. His analysis of the meaning of wrestling as a performance in which the event on stage, and the audience watching, collaborate in a sophisticated exchange of moral adjudication, points to the subtle and sometimes overt ways in which the seemingly 'sporty' activity of professional wrestling represents a complex semiotic triangulation between the idea, the performance and its consumption. Through the making of moral meaning wrestling becomes a hermeneutic enactment. Similarly, in Ireland staging the recruitment of soldiers and the remembrance of the dead engaged the population in a complex interpretative exercise. What took place was a dialogue between remembering and forgetting, between providing moral legitimacy or denying it.

The representation of the war in recruitment posters in Ireland simultaneously focused on the unique and the universal by deploying tropes of local landscape identity to entice men to enlist while also making reference to more universal images of a 'just' cause. These images were further mediated through pamphlets circulated by politicians, clergy and military leaders. Daniels and Cosgrove rightly claim that 'Spectacle and text, image and word have always been dialectically related, not least in theatre itself, and this unity has been the site of an intense struggle for meaning.'[2] Throughout the discussions of different stages of memory in this book, the interrelationships between the word, the public performance of remembrance and the aesthetics of collective ritual have been emphasised. I have sought to stress that visual representation alone masks some of the deeper fissures that have informed the public performances of social memory. The parading of Peace Day celebrations in Dublin and other smaller towns across the country were profoundly bound up with a discussion of location both in the spatial sense of the routes of parades through the streets and also in relation to the location of the war in Ireland's history. One representative icon of war – the soldier – was differentiated between the war veteran in civilian clothing and the professional soldier in military uniform. While one could be celebrated as a representation of bravery at the war front, the other could be seen as a representation of oppression at home. Public hostility to such iconography was periodically expressed in the violent reaction to uniformed soldiers occupying public space. The distinction between war abroad and peace at home was subverted in the Irish case where the struggle for political independence had become both a verbal and a physical conflict. By way of contrast,

---

[2] Daniels and Cosgrove, 'Spectacle and text', 59.

in Belfast and in other northern towns, such subversion did not take place and the celebration of victory was literally the celebration of the defence of British soil. The larger scale of public performance and the selection of August in which to host the ceremonies reinforced the notion that Ulster was different from much of the remainder of Ireland. The fact that it located memory in a different time marked out this exceptionalism.

While parades may act as fleeting, albeit repetitive, signifiers of remembrance, public memorials are permanent markers of the war. The poetics and politics of representation reiterate the role of the spatial and the allegorical in the contest over public memorials. The mapping of Dublin's memory sites was directly implicated in debates surrounding the national memorial. Parliamentary discourse on this memorial revolved around a number of issues, but the discussions on the location of the memorial park were influenced by the connections being drawn between the micro-geography of the site and broader historiographical questions. The clear parallel of historical interpretation with spatial setting indicated that the drama of the First World War could not be rehearsed adjacent to the drama of the Easter Rising and the political centre of the state for fear of interpretative misunderstanding. The aesthetics of the memorial to the war dead, then, was not an aesthetics of artistic design *per se* but one of aestheticised location. The 1916 rebellion could be celebrated in the heart of the capital and its epicentre reflected the strategic heart of the rebellion itself. In other words, the parades and memorial to the Rising were directly mapped on to the geography of the conflict and the intellectual dramatisation of the rebellion by its leaders as an exercise in national martyrdom was literally and symbolically reinforced by this action.

Geography became central to the manner in which meaning would be conveyed and those who organised memorials and parades were mindful of this in their planning. The setting did not merely act as a backdrop to the theatricality of public remembrance but it was central to the construction of meaning. While Jay Winter has reminded us of the importance of mourning in the dialectics of remembrance for those commemorating casualties of the Great War, personal memory was much more remote in the context of the 1916 Rising, as far fewer families were directly touched by bereavement. Paradoxically, the translation of the 1916 Rebellion – from a comparatively small skirmish into an episode of national sacrifice and national mourning – ensured that its entry into the public imagination was powerfully and easily achieved. The dramatic context of the Rising prompted its relentless inscription through literature, although cynical and ironic responses to the conflict, which emphasised its futility, were far rarer than in the case of the Great War. Some literary texts of the First World War had abandoned the high diction of previous generations in favour of modernist forms of representation to react against the morality of the conflict. Where the Rising is concerned modernist and abstract forms of representation could easily coexist, without fear of contradiction, with a celebratory mode of remembrance. Abstraction worked in the context of the Rising without eliciting cynicism, as the literal spaces of the revolution had superimposed on them figures who inhabited the mythological world – Cuchulain and the children of Lír – and embodied a sense of collective identity irreducible to individual, historical, rebels.

The translation of the front of the First World War to the Home Front through literary texts sought to bridge the gap – both spatial and allegorical – between the imagination of those engulfed in the conflict in the war zones and those who remained at home. A tension between these two spaces pervades the writings of Irish authors. In O'Casey's work the deployment of a Christian imaginary, and of the Catholic Mass as an organising motif for dramatising the war for home consumption, is especially visible. For the writer the challenge to implicate the non-combatant population in the exercise of war, and consequently to compel them not to wilfully forget and absolve themselves of all responsibility, accounts perhaps for the initial hostility towards the play. The experimental style of the spectacle presents the audience with that necessity for moral judgement that animates Barthes' discussion of wrestling. Not happy to allow those on the Home Front to distance themselves from the brutality of the action at the war front, the play perhaps addresses more widely the question of allowing Irish society in general to cajole itself into thinking that it had fought somebody else's war. The writings of other less distinguished authors than O'Casey replicate, to a degree, this uneasy position of scripting and representing the war from the viewpoint of the soldier stationed along the front, yet remaining cognisant of the potential antipathy of those occupying lands remote from that battle zone.

In the shadow of all efforts to stage remembrance of Ireland's war casualties was the ongoing political movement at home and specifically the Rising of 1916. This book has consistently stressed the 'civil war' of identity and memory that underpinned much of the discussion surrounding this process. That civil strife found expression regionally in the differential commitment to memory work and locally in the choice, nomenclature and iconography associated with particular sites. Social memory is never a simple empty space awaiting manipulation by the powerful. Rather it is a messy space where competing and at times conflicting memories are accumulated, accreted, refined and sometimes challenged. That a relatively small guerrilla conflict, where the weak were overcome by the strong, could capture the public imagination more easily and more powerfully than a long protracted war in Europe whose aims were misted through the passage of time is not altogether surprising. Making sense of the war by all participating states entailed focusing through a national lens and a discourse that could be domesticated to accommodate national ideologies and circumstances. In Ireland that concept of 'the national' was fractured along fault lines of class, religion and political identity and those fault lines were not just contained within the mental maps of Ireland's citizenry but were played out in the spaces where those memories were materially inscribed. More broadly, then, I wish to suggest that this story illuminates the significance of the spatiality of memory to discussions of the cultural meaning of war. Space is more than the container in which historical narratives of memory are placed. Our attention is thus re-directed towards the manner in which the landscapes of war's social memory – the texts, theatres, townscapes – become the process of memory construction rather than its outcome. That this process involves inscription and erasure, consensus and conflict, joy and pain, reflection and action speaks towards the dilemma that warfare represents in the public performance of remembrance.

Bibliography

Agulhon, M., *Marianne into battle: republican imagery and symbolism in France, 1789–1880* (Cambridge: Cambridge University Press, 1981).

Allen, J. L., ed., 'Creation of myth: invention of tradition in America', *Journal of Historical Geography* (1992) 18, 1–138.

An open letter from Capt. Stephen Gwynn MP to the young men of Ireland, n.d. (Trinity College Dublin: Recruiting leaflets relating to European War, 1914–18, OLS L-1-540 Nos. 1–16).

Anderson, B., *Imagined communities: on the origins and spread of nationalism* (London: Verso, 1989).

Armstrong, K. and Benyon, H., eds., *Hello are you working?! Memories of the thirties in the north east of England* (Durham: Strong Words, 1977).

Atkinson, D. and Cosgrove, D., 'Urban rhetoric and embodied identities: city, nation and empire at the Vittorio Emanuele II monument in Rome, 1870–1945', *Annals of the Association of American Geographers*, 88 (1998), 28–49.

Auster, M., 'Monument in a landscape: the question of "meaning"', *Australian Geographer*, 28 (1997), 219–27.

Azaryahu, M., 'From remains to relics: authentic monuments in the Israeli landscape', *History and Memory*, 5 (1993), 82–103.

B.-W., J. 'The Silver Tassie', *New Statesman*, 3 October 1929.

Baker, P., *King and country call: New Zelanders, conscription and the Great War* (Auckland: Auckland University Press, 1988).

Barker, P., *Regeneration* (London: Penguin, 1991).

Barnes, T. and Duncan, J., eds., *Writing worlds: discourse, text and metaphor in the representation of landscapes* (London: Routledge, 1992).

Barry, J., *A short history of the famous Catholic necropolis* (Dublin: Dolmen, 1932).

Barthes, R., *The elements of semiology* (London: Cape, 1967).

'The world of wrestling' in *Mythologies*, trans. Annette Lavers (New York: Hill and Wang, 1972), 18–30.

Mythologies (London: Cape, 1972).

The pleasure of the text (New York: Hall and Wang, 1975).

Image, music text (New York: Hill and Wang, 1987).

'The advertising message' in R. Barthes, *The semiotic challenge* (Oxford: Basil Black-well), 1988, 173–8.

Bartlett, T. and Jeffery, K., eds., *A military history of Ireland* (Cambridge: Cambridge University Press, 1996).

Becker, A., *Les monuments aux morts: mémoire de la Grande Guerre* (Paris: Errance, 1988).

Bell, J., 'Redefining national identity in Uzbekistan: symbolic tensions in Tashkent's official public landscape', *Ecumene*, 6 (1999), 183–213.

Bodnar, J., *Remaking America: public memory, commemoration and patriotism in the twentieth century* (Princeton: Princeton University Press, 1992).

Bolger, D., ed., *Francis Ledwidge: selected poems* (Dublin: New Island Books, 1992).

Bonadeo, A., *Mark of the beast: death and degradation in the literature of the Great War* (Kentucky: University of Kentucky Press, 1989).

Bonnett, A., 'Situationism, geography and poststructuralism', *Environment and Planning D: Society and Space*, 7 (1989), 131–46.

Bowman, T., 'Composing divisions', *Causeway*, 2 (1995), 24–9.

Boyarin, J., *Remapping memory: the politics of timespace* (London: University of Minnesota Press, 1994).

Boyce, G., *The sure confusing drum: Ireland and the First World War* (Swansea: University of Wales Press, 1993).

'Ireland and the First World War', *History Ireland*, 2 (1994), 48–53.

Boyle, J. W., *Leaders and workers* (Cork: Mercier, 1966).

Bracco, R. M., *Merchants of hope: British middlebrow writers and the First World War, 1919–39* (Oxford: Berg, 1993).

Brady, C., ed., *Interpreting Irish history: the debate on historical revisionism* (Dublin: Irish Academic Press, 1994).

Bravo, A., 'Italian peasant women and the First World War' in C. Emsley, A. Marwick and W. Simpson, eds., *War, peace and social change in twentieth century Europe* (Milton Keynes: Open University Press, 1989), 102–15.

Braybon, C., *Women workers in the First World War: the British experience* (London: Croom Helm, 1981).

Bredin, A. E. C., *History of the Irish soldier* (Belfast: Century Books, 1987).

British Legion, *Victory Souvenir* (Dublin: J. J. McCann and Co., 1946).

British Legion Annual, *Irish National War Memorial* (Dublin: Alex Thom, 1941).

Burke, J., *Dismembering the male: men's bodies, Britain and the Great War* (London: Reaktion Books, 1996).

Bushaway, B., 'Name upon name: the Great War and remembrance' in R. Porter, ed., *Myths of the English* (Cambridge: Polity, 1992), 136–67.

Butler, J., *Gender trouble: feminism and the subversion of identity* (London: Routledge, 1990).

Callan, P., 'Recruiting for the British army in Ireland during the First World War', *Irish Sword*, 17: 67 (1987), 42–54.

Campbell, D., *Women at war with America: private lives in a patriotic era* (Cambridge, MA: Harvard University Press, 1984).

Carruthers, M., *The book of memory: a study of memory in medieval culture* (Cambridge: Cambridge University Press, 1990).

Central Council for Recruiting in Ireland, *Pamphlet* (Trinity College Dublin: Recruiting leaflets relating to European War, 1914–18, OLS L-1-540 Nos. 1–16), n.d.

Charlesworth, A., 'Contesting places of memory: the case of Auschwitz', *Environment and Planning D: Society and Space*, 12 (1994), 579–93.

Clout, H., *After the ruins: restoring the countryside of Northern France after the Great War* (Exeter: University of Exeter Press, 1996).

Collingwood, R., *The idea of history* (Oxford: Oxford University Press, 1946).

Connerton, P., *How societies remember* (Cambridge: Cambridge University Press, 1989).

Connolly, M., 'James Connolly: socialist and patriot', *Studies*, 41 (1952), 293–308.

Cooke, S., 'Negotiating memory and identity: the Hyde Park Holocaust Memorial, London', *Journal of Historical Geography*, 26 (2000), 449–65.

Cosgrove, D., *Social formation and symbolic landscape* (London: Croom Helm, 1984).

The Palladian landscape: geographical change and its cultural representations in sixteenth century Italy (University Park: Pennsylvania State University Press, 1993).

Cosgrove, D. and Daniels, S., eds., *The iconography of landscape* (Cambridge: Cambridge University Press, 1988).

Cowasjee, S., *Sean O'Casey, The man behind the plays* (London: Oliver and Boyd, 1963).

Craig, M., *Dublin 1660–1860* (Dublin: Allen Figgis, 1980).

Cunliffe, M., *The Royal Irish Fusiliers, 1793–1950* (Oxford: Oxford University Press, 1971).

Curtayne, A., *Francis Ledwidge: a life of the poet* (London: Martin, Brian and O'Keeffe, 1972).

Curtis, L. P., 'Ireland in 1914' in W. E. Vaughan, ed., *A new history of Ireland* (Oxford: Clarendon, 1996), 175–88.

Apes and angels: the Irishman in Victorian caricature (Newton Abbot: David and Charles, 1971).

Dáil Éireann, *Official Report*, xix, 29 March 1927.

Daly, M., *Dublin: The deposed capital. A social and economic history 1860–1914* (Cork: Cork University Press, 1984).

Daniels, S. and Cosgrove, D., 'Spectacle and text: landscape metaphors in cultural geography' in J. Duncan and D. Ley, eds., *Place/culture/representation* (London: Routledge, 1993), 57–77.

Darracott, J., *The First World War in posters* (London: Constable, 1974).

Davis, S., 'Empty eyes, marble hand: the Confederate monument and the South', *Journal of Popular Culture*, 16 (1982), 2–21.

De Saussure, F., *Course in general linguistics* (London: Peter Owen, 1960).

Denman, T., 'Sir Lawrence Parsons and the raising of the 16th (Irish) Division, 1914–15', *Irish Sword*, 17: 67 (1987), 90–104.

'The 10th (Irish) Division 1914–15: A study in military and political interaction', *Irish Sword*, 17:66 (1987), 16–25.

'The Catholic Irish soldier in the First World War: the "racial" environment', *Irish Historical Studies*, 108 (1991), 352–65.

Ireland's unknown soldiers: the 16th Irish Division (Dublin: Irish Academic Press, 1992).

'Irish politics and the British army list: the formation of the Irish Guards in 1900', *Irish Sword*, 19:77 (1995), 171–86.

Department of Early Books, Trinity College Dublin (Call nos. 22.7.27–29, Papyrus Case 16).

Doherty, G., 'Post-Famine emigration' in S. Duffy, ed., *Atlas of Irish history* (Dublin: Gill and Macmillan, 1997), 102–3.

Dombrowski, N., *Women and war in the twentieth century: enlisted with or without consent* (London: Garland, 1999).

Dooley, T., 'Politics, bands and marketing: army recruitment in Waterford city, 1914–15', *Irish Sword*, 18 (1991), 205–19.

Dudley Edwards, O., 'Patrick MacGill and the making of a historical source: with a handlist of his works', *The Innes Review*, 37 (1986), 73–99.

Duncan, J. S., *The city as text: the politics of landscape interpretation in the Kandyan Kingdom* (Cambridge: Cambridge University Press, 1990).

Duncan, J. and Duncan, N., '(Re)reading the landscape', *Environment and Planning D: Society and Space*, 6 (1988), 117–26.

'Ideology and bliss: Roland Barthes and the secret histories of landscape' in T. Barnes and J. S. Duncan, *Writing worlds: discourse, text and metaphor in the representation of landscape* (London: Routledge, 1992), 18–37.

Duncan, J. S. and Ley, D., eds., *Place/Culture/Representation* (London: Routledge, 1993).

Dungan, M., *Distant drums: Irish soldiers in foreign armies* (Belfast: Appletree, 1993).

Irish voices from the Great War (Dublin: Irish Academic Press, 1995).

Dunsany, Lord, ed., *The complete poems of Francis Ledwidge* (London: Herbert Jenkins, 1919).

Eagleton, T., *Literary theory: an introduction* (Minneapolis: University of Minnesota Press, 1983).

Eksteins, M., *Rites of spring: the Great War and the birth of the modern age* (New York: Bantam Books, 1989).

Ellis, J., '"The methods of barbarism" and the "Rights of small nations": war propaganda and British pluralism', *Albion*, 37 (1998), 49–75.

'The degenerate and the martyr: nationalist propaganda and the contestation of Irishness. 1914–1918', *Eire-Ireland*, 35 (2000), 7–33.

Ellis-Fermor, U., *The frontiers of drama* (London: Methuen, 1945).

Elshtain, J., *Women and war* (Brighton: Harvester, 1987).

Englander, D., 'Soldiering and identity: reflections on the Great War', *War in History*, 1 (1994), 300–18.

English, R. and Walker, G., eds., *Unionism in modern Ireland: new perspectives on politics and culture* (Dublin: Gill and Macmillan, 1996).

Fairhall, J., *James Joyce and the question of history* (Cambridge: Cambridge University Press, 1993).

Faulks, S., *Birdsong* (London: Hutchinson, 1993).

Fitzpatrick, D., 'Militarism in Ireland 1900–1922' in T. Bartlett and K. Jeffery, eds., *A military history of Ireland* (Cambridge: Cambridge University Press, 1996), 379–406.

'Commemoration in the Irish Free State: a chronicle of embarrassment' in I. McBride, ed., *History and memory in modern Ireland* (Cambridge: Cambridge University Press, 2001), 184–203.

Fitzpatrick, D., ed., *Ireland and the First World War* (Dublin: Trinity College Workshop, 1986).

Fitzpatrick, W. J. F., *History of some Dublin Catholic cemeteries* (Dublin: Macmillan, 1900).

Foster, G. M., *Ghosts of the confederacy: defeat, the lost cause, and the emergence of the new South* (Oxford: Oxford University Press, 1987).

Foster, J. W., *Forces and themes in Ulster fiction* (Dublin: Gill and Macmillan, 1974).

'Imagining the *Titanic*' in E. Patten, ed., *Returning to ourselves* (Belfast: Lagan Press, 1995), 325–43.

Foster, R., *Modern Ireland 1600–1972* (London: Allen Lane, 1988).

Paddy and Mr. Punch: connections in Irish and English history (London: Allen Lane, 1993).

Foucault, M., *Power/knowledge* (Brighton: Harvester, 1980).

Friel, B., *Translations* (London: Faber and Faber, 1981).

Fussell, P., *The Great War and modern memory* (Cambridge: Cambridge University Press, 1975).

Gaffney, A., *Aftermath: remembering the Great War in Wales* (Cardiff: University of Wales Press, 1998).

Garvin, T., *The evolution of Irish nationalist politics* (Dublin: Gill and Macmillan, 1983).

Gassner, J., *Masters of drama* (New York: Random House, 1954).

Gibbons, L., *Transformations in Irish culture* (Cork: Cork University Press, 1996).

Gilbert, S. M. and Gubar, S., *No man's land: The place of the woman writer in the twentieth century. Vol. 2: Sexchanges* (New Haven: Yale University Press, 1989).

Gillis, J. R., 'Memory and identity: The history of a relationship' in R. Gillis, ed., *Commemorations: the politics of national identity* (Princeton: Princeton University Press, 1994), 3–24.

Gradidge, R., *Edwin Lutyens: architect laureate* (London: Allen and Unwin, 1981).

Graham, G., *The shape of the past* (Cambridge: Cambridge University Press, 1997).

Granville-Barker, H., *On poetry in drama* (London: The Romance Lecture, 1937).

Gregory, A., *Lady Gregory's journals 1916–30*, ed., L. Robinson (London: Putnam, 1946).

The silence of memory: Armistice Day 1919–1946 (Oxford: Berg, 1994).

Gulley, H. E., 'Women and the lost cause: preserving Confederate identity in the American Deep South', *Journal of Historical Geography*, 19 (1993),125–41.

Gwynn, D., *The life of John Redmond* (London: Harrap, 1932).

Halbwachs, M., *On collective memory*, L. Coser, ed. and trans. (Chicago: Harper Row, 1992). Originally published in French as *La mémoire collective* (Paris, 1950).

Hanley, L., *Writing war: fiction, gender and memory* (Amherst: University of Massachusetts Press, 1991).

Harbison, P., Potterton, H. and Sheehy, J., *Irish art and architecture from prehistory to the present.* (London: Thames and Hudson, 1978).

Hardie, M. and Sabin, A. K., *War posters* (London: A. & C. Black, 1920).

Harley, B., 'Deconstructing the map' in T. Barnes and J. Duncan, eds., *Writing worlds: discourse, text and metaphor in the representation of landscape* (London: Routledge, 1991), 231–47.

Harris, H. E., *The Irish regiments in the First World War* (Dublin: Mercier, 1968).

Harvey, D., 'Monument and myth', *Annals of the Association of American Geographers*, 69 (1979), 362–81.

Haste, C., *Keep the home fires burning: propaganda in the First World War* (London: Allen Lane, 1977).

Heffernan, M., 'For ever England: the Western Front and the politics of remembrance in Britain', *Ecumene*, 2 (1995), 293–324.

Higonnet, M. and Higonnet, P. L.-R., 'The double helix' in M. Higonnet *et al.*, eds., *Behind the lines: gender and two world wars* (New Haven: Yale University Press, 1987), 31–50.

Higonnet, M., Jenson, J., Michel, S. and Weitz, M., eds., *Behind the lines: gender and two world wars* (London: Yale University Press, 1987).

Hillier, B., *Posters* (Hamlyn: Feltham, 1969).

Hobsbawm, E. and Ranger, T., eds., *The invention of tradition* (Cambridge: Cambridge University Press, 1983).

Howie, D. and Howie, J., 'Irish recruiting and the Home Rule crisis of August-September 1914' in M. Dockrill and D. French, eds., *Strategy and intelligence: British policy during the First World War* (London: Hambledon Press, 1996), 1–22.

Hutchinson, J., *The dynamics of cultural nationalism: the Gaelic Revival and the creation of the Irish nation state* (London: Allen and Unwin, 1987).

Hutton, P., *History as an art of memory* (Burlington, VT: University of New England Press, 1993).

Hynes, S., *A war imagined: the Great War and English literature* (London: Bodley Head, 1991).

'Personal narratives and commemoration' in J. Winter and E. Sivan, eds., *War and remembrance in the twentieth century* (Cambridge: Cambridge University Press, 1999), 205–20.

Inglis, K. S., *Sacred places: war memorials in the Australian landscape* (Melbourne: University of Melbourne Press, 1998).

Inglis, K. S. and Phillips, J.,'War memorials in Australia and New Zealand: a comparative survey', *Australian Historical Studies*, 24 (1991), 171–91.

Irving, R. G., *Indian summer: Lutyens, Baker and Imperial Delhi* (London: Yale University Press, 1981).

Jackson, J. L., *French patriotic posters 1914–1920* (Rice University, Houston, TX: Sewall Art Gallery Exhibition Catalogue, 1989).

Jackson, P., *Maps of meaning* (London: Routledge, 1989).

Jameson, F., *The political unconscious* (Ithaca: Cornell University Press, 1981).

Jarman, N., *Material conflicts: parades and visual displays in Northern Ireland* (Oxford: Berg, 1997).

Jay, M., 'In the empire of the gaze: Foucault and the denigration of vision in twentieth century French thought' in D. C. Hoy, ed., *Foucault: a critical reader* (Oxford: Blackwell, 1986), 175–204.

Jeffery, K. 'The Great War and modern Irish memory' in T. G. Fraser and K. Jeffery, eds., *Men, women and war* (Dublin: Lilliput Press, 1993), 136–57.

'Irish artists and the First World War', *History Ireland*, 1 (1993), 42–45.

'Irish culture and the Great War', *Bullán*, 1 (1994), 87–96.

Ireland and the Great War (Cambridge: Cambridge University Press, 2000).

Jeffery, K., ed., *Men, women and war* (Dublin: Irish Academic Press, 1993).

Jennings, R., 'The Silver Tassie by Sean O'Casey: at the Apollo Theatre', *Spectator*, 143 (October 1929).

Johnson, N. C., 'Sculpting heroic histories: celebrating the centenary of the 1798 rebellion in Ireland', *Transactions of the Institute of British Geographers*, 19 (1994), 78–93.

'Cast in stone: monuments, geography and nationalism', *Environment and Planning D: Society and Space*, 13 (1995), 51–65.

Johnson, R., McLennan, G., Schwarz, B. and Sutton, D., eds., *Making histories: studies in history-writing and politics* (London: Hutchinson, 1982).

Johnstone, T., *Orange, green and khaki: the story of Irish regiments in the Great War, 1914–18* (Dublin: Gill and Macmillan, 1992).

Jourdain, H. F. N., *History of the Connaught Rangers*, 3 vols. (London: Royal United Service Institution, 1925–28).

Joyce, J., 'The dead' in *Dubliners* (London: Paladin, 1988). Originally published in 1914 by Maunsel and Co.

Ulysses, edited with introduction by Jeri Johnson (Oxford: Oxford University Press, 1993). Originally published in 1922.

Kedar, B. and Werblowsky, R., eds., *Sacred space: shrine, city, land* (New York: New York University Press, 1998).

Kerr, S. P., *What the Irish regiments have done* (London: Unwin, 1916).

Kertzer, D., *Ritual, politics and power* (London: Yale University Press, 1988).

Kiberd, D., 'The elephant of revolutionary forgetfulness' in M. Ní Dhonnchadha and T. Dorgan, eds., *Revising the rising* (Derry: Field Day, 1991), 1–20.

Inventing Ireland: the literature of the modern nation (London: Jonathan Cape, 1995).

King, A., *Memorials of the Great War in Britain* (Oxford: Berg, 1998).

Klaus, H. G., *The socialist novel in Britain* (Brighton: Harvester, 1982).

Kleiman, C., *Sean O'Casey's bridge of vision* (London: University of Toronto Press, 1982).

Krell, D., *Of memory, reminiscence and writing* (Bloomington: University of Indiana Press, 1990).

Lacqueur, T. W., 'Memory and naming in the Great War' in J. R. Gillis, ed., *Commemorations: the politics of national identity* (Princeton: Princeton University Press, 1994), 150–67.

Langbein, H., 'The controversy over the convent at Auschwitz' in C. Rittner and J. K. Roth, eds., *Memory offended: the Auschwitz convent controversy* (New York: Praeger, 1991), 95–8.

Larkin, E., *James Larkin: Irish labour leader, 1876–1947* (London: Routledge and Keegan, 1965).

Larmour, P., *Belfast: an illustrated architectural guide* (Belfast: Friar's Bush Press, 1987).

Law, C., *Suffrage and power: the women's movement, 1918–28* (London: I.B. Tauris, 1997).

Le Goff, J., *History and memory*, trans. S. Rendall and E. Clamen (New York: Albany Press, 1992).

Lee, J. J., *Ireland 1912–1985: Politics and society* (Cambridge: Cambridge University Press, 1989).

Leed, E., *No man's land: combat and identity in World War One* (Cambridge: Cambridge University Press, 1979).

Leerssen, J., *Remembrance and imagination: patterns in the historical and literary representation of Ireland in the nineteenth century* (Cork: Cork University Press, 1996).

Lefebvre, H., *The production of space* (Oxford: Blackwell, 1991).

Leib, J. I., 'Separate times, shared spaces: Arthur Ashe, Monument Avenue and the politics of Richmond, Virginia's symbolic landscape', *Cultural Geographies*, 9 (2002), 286–312.

Leitner, H. and Kang, P., 'Contested urban landscapes of nationalism: the case of Taipei', *Ecumene*, 6 (1999), 172–92.

Leonard, H., 'Aldwych: *The Silver Tassie*', *plays and players* (November 1969), 20–3.

Leonard, J., 'Lest we forget' in D. Fitzpatrick, *Ireland and the First World War* (Dublin, Trinity History Workshop, 1986), 59–67.

'The twinge of memory: Armistice Day and Remembrance Sunday in Dublin since 1919' in R. English and G. Walker, eds., *Unionism in modern Ireland* (Dublin: Gill and Macmillan, 1996), 99–114.

Levenson, M., *A genealogy of modernism: a study of English literary doctrine 1908–1922* (Cambridge: Cambridge University Press, 1984).

Ley, D. and Olds, K., 'Landscape as spectacle: world's fairs and the culture of heroic consumption', *Environment and Planning D: Society and Space*, 6 (1988), 191–212.

Liddy, P., *Dublin be proud* (Dublin: Chadworth, 1987).

Lipsitz, G., *Time passages: collective memory and American popular culture* (Minneapolis: University of Minneapolis Press, 1990).

Lloyd, D., *Anomalous states: Irish writing and the postcolonial moment* (Dublin: Lilliput Press, 1993).

Lyons, F. S. L., *Ireland since the famine* (London: Weidenfeld and Nicholson, 1971).

'The revolution in train' in W. E. Vaughan, ed., *A new history of Ireland* (Oxford: Clarendon), vol. VI, 1996, 189–206.

Mac Aonghusa, P. and Ó Réagáin, L., eds., *The best of Pearse* (Cork: Mercier Press, 1967).

Mac Donagh, M., *The Irish at the Front* (London: Hodder and Stoughton, 1916).

MacDonnel, A. G., 'Chronicles, the drama', *London Mercury*, December 1929.

MacGill, P., *The amateur army* (London: Herbert Jenkins, 1915).

The rat-pit (London: Herbert Jenkins, 1915).

The red horizon (London: Herbert Jenkins, 1916).

The great push (London: Herbert Jenkins, 1916). Reprinted by Caliban Books, 1984.

MacGreevy, T., *Collected Poems*, ed., Thomas Dillon Redshaw (Dublin: New Writers Press, 1971).

Maddocks, F., 'Turnage scores in injury time', *The Observer*, 20 February 2000.

Mansergh, N., *The Irish question 1840–1921* (London: Unwin, 1965).

Martin, F. X., *The Irish Volunteers, 1913–15* (Dublin: Duffy, 1963).

Leaders and men of the Easter Rising: Dublin 1916 (London: Methuen, 1967).

Martin, F. X. and Byrne, F. J., eds., *The scholar revolutionary: Eoin MacNeill, 1867–1945 and the making of the new Ireland* (New York: Harper and Row, 1973).

McBride, I., ed., *History and memory in modern Ireland* (Cambridge: Cambridge University Press, 2001).

McGuinness, F., *Observe the sons of Ulster marching towards the Somme* (London: Faber and Faber, 1986).

McKinnion, M., *New Zealand historic atlas* (Auckland: Bateman, 1997).

McMullan, Reverend J., CP, *Irish soldiers at the front* (Trinity College Dublin: Recruiting leaflets relating to European War, 1914–18, OLS L-1-540 Nos. 1–16), n.d.

Meinig, D., ed., *The interpretation of ordinary landscapes* (Oxford: Oxford University Press, 1979).

Middleton, D. and Edwards, D., eds., *Collective remembering* (London: Sage, 1990).

Mitchell, D., *Cultural geography: a critical introduction* (Oxford: Blackwell, 2000).

Moriarty, C., 'Christian iconography and First World War memorials', *Imperial War Museum Review*, 6 (1990), 63–75.

Morris, M. S., 'Gardens "Forever England": landscape, identity and the First World War cemeteries on the Western Front', *Ecumene*, 4 (1997), 410–34.

Mossé, G., *The nationalization of the masses* (New York: Howard Fertig, 1975).

Fallen soldiers: shaping the memory of two world wars (Oxford: Oxford University Press, 1990).

Nash, C., 'Embodying the nation: the west of Ireland landscape and Irish identity' in B. O'Connor and M. Cronin, eds., *Tourism Ireland: a critical analysis* (Cork: Cork University Press, 1993), 86–112.

'Renaming and remapping', *Feminist Review*, 44 (1993), 39–57.

'Remapping the body/land: new cartographies of identity, gender and landscape in Ireland' in A. Blunt and G. Rose, eds., *Writing women and space: colonial and postcolonial geographies* (New York: Guilford, 1994), 227–50.

National Archive of Ireland, *Department of the Taoiseach* (DT) S4156A, 'Proposal outlined at meeting of cabinet by Mr. T. J. Byrne, Principal Architect, Board of Works', 29 October 1929.

National Gallery of Ireland Catalogue, *Golden jubilee of the Easter Rising* (Dublin: Dolmen Press, 1966).

National Graves Association, *The Last Post: Glasnevin cemetery being a record of Ireland's heroic dead in Dublin city and county. Also places of historic interest* (Dublin: National Graves Association, 1932). Compiled and edited by Mary Donnelly.

Souvenir programme, Mansion House, 23 November 1937.

Manchester martyrs commemoration concert programme, Mansion House, 21 November 1941.

Waterford remembers (Waterford: Waterford News Limited, 1946). Written by N. de Fuiteóil.

The Last Post, 2nd edn (Dublin: National Graves Association, 1976).

Nicholls, P., *Modernisms: a literary guide* (Basingstoke: Macmillan, 1995).

Nicoll, A., *British drama* (London: Harrap, 1949).

Nora, P., 'Between memory and history: les lieux de mémoire', *Representations*, 26 (1989), 7–25.

Nora, P., ed., *Realms of memory: Vol. 11: traditions* (Chichester: Columbia University Press, 1997).

Ó Cuív, B., ed., *A view of the Irish language* (Dublin: Stationery Office, 1969).

Ó hÓgáin, D., *Myths, legend and romance: an encyclopedia of the Irish folk tradition* (London: Ryan, 1990).

Ó Tuama, S., ed., *The Gaelic League idea* (Dublin: Mercier Press, 1972).

O'Brien, C. C., *States of Ireland* (London: Hutchinson, 1972).

O'Brien, J. H., *Liam O'Flaherty* (Lewisburg: Bucknell University Press, 1973).

O'Broin, L., *Revolutionary underground: the story of the Irish Republican Brotherhood 1858–1924* (Dublin: Gill and Macmillan, 1976).

O'Casey, S., *Mirrors in my house: the autobiographies of Sean O'Casey* (New York: Macmillan, 1956).

'The Silver Tassie' in *Seven plays by Sean O'Casey*, Selected and introduced by R. Ayling (London: Macmillan, 1985). Originally published in 1928.

O'Flaherty, L., *The Return of the brute* (Dublin: Wolfhound Press, 1998). Originally published 1929.

O'Keefe, T. J., 'The 1898 efforts to celebrate the United Irishmen: the '98 centennial', *Éire-Ireland*, 23 (1988), 51–73.

' "Who fears to speak of '98": the rhetoric and rituals of the United Irishmen centennial, 1898', *Éire-Ireland*, 28 (1992), 67–91.

O'Sullivan, P., 'Patrick MacGill: the making of a writer' in S. Hutton and P. Steward, eds., *Ireland's histories: aspects of state, society and ideology* (London: Routledge, 1991), 203–22.

Orr, P., *The road to the Somme* (Belfast: Blackstaff, 1987).

Osborne, B., 'The iconography of nationhood in Canadian art' in D. Cosgrove and S. Daniels, eds., *The iconography of landscape* (Cambridge: Cambridge University Press, 1988), 162–78.

'Figuring space, marking time: contested identities in Canada', *International Journal of Heritage Studies*, 2 (1996), 23–40.

'Warscapes, landscapes, inscapes: France, war, and Canadian national identity' in I. Black and R. Butlin, eds., *Place, culture and identity* (Quebec: Les Presses de l'Université Laval, 2001), 311–33.

Ouditt, S. *Fighting forces, writing women: identity and ideology in the First World War* (London: Routledge, 1994).

Ozouf, M., *Festivals and the French Revolution* (Cambridge, MA: Harvard University Press, 1988).

Peet, R., 'A sign taken from history: Daniel Shay's memorial in Petersham, Massachusetts', *Annals of the Association of American Geographers*, 86 (1996), 21–43.

Perry, N., 'Nationality in the Irish infantry regiments in the First World War', *War and Society*, 12 (1994), 65–95.

Phelan, K., 'A note on O'Casey', *Commonweal*, L, 7 October 1949.

Pickles, J., 'Texts, hermeneutics and propaganda maps' in T. Barnes and J. Duncan, eds., *Writing worlds: discourse, text and metaphor in the representation of landscape* (London: Routledge, 1992), 193–230.

Piehler, G. K., 'The war dead and the gold star: American commemoration of the First World War' in J. R. Gillis, ed., *Commemorations: the politics of national identity* (Princeton: Princeton University Press, 1994), 168–85.

Pierce, J., 'Constructing memory: the Vimy memorial', *Canadian Military History*, 1 (1992), 1–3.

Porter, R., ed., *Myths of the English* (Cambridge: Polity, 1992).

Prost, A., 'Monuments to the dead' in P. Nora, ed., *Realms of memory, Vol. II: traditions* (Chichester: University of Columbia Press, 1992), 307–32.

Pugh, M., *Women's suffrage in Britain, 1867–1928* (London: Historical Association, 1980).

Rickards, M., *Posters of the First World War* (London: Evelyn, Adams and MacKay, 1968).

Riley, D., *War in the nursery* (London: Virago, 1983).

Rollins, R. G., *Sean O'Casey's drama: verisimilitude and vision* (Alabama: University of Alabama Press, 1979).

Rover, C., *Women's suffrage and party politics in Britain, 1866–1914* (London: Routledge and Keegan, 1967).

Ryan, J., *Picturing empire* (London: Reaktion, 1997).

Said, E., *Orientalism* (London: Vintage, 1979).

Samuel, R., *Theatres of memory*, vol. I (London: Verso, 1994).

Sassoon, S., 'The glory of women' in I. M. Parsons, ed., *Men who march away: poems of the First World War* (London: Chatto and Windus, 1965).

Savage, K., 'The politics of memory: black emancipation and the Civil War monument', in R. Gillis, ed., *Commemorations: the politics of national identity* (Princeton: Princeton University Press, 1994), 127–49.

Schein, R., 'The place of landscape', *Annals of the Association of American Geographers*, 87 (1997), 660–80.

Schorske, C.E., *Fin-de-siècle Vienna: politics and culture* (London: Weidenfeld and Nicolson, 1979).

Seanad Éireann, *Official Report*, viii, 9 March 1927.

Sheehy, J., *The rediscovery of Ireland's past: the Celtic revival 1830–1930* (London: Thames and Hudson, 1980).

Sheeran, P. F., *The novels of Liam O'Flaherty: a study in romantic realism* (Dublin: Wolfhound Press, 1976).

Sherman, D., *The construction of memory in interwar France* (London: University of Chicago Press, 1999).

'Art, commerce and the production of memory in France after World War I' in J. R. Gillis, ed., *Commemorations: the politics of national identity* (Princeton: Princeton University Press, 1994), 186–211.

Sherry, R., 'The Irish working class in fiction' in J. Hawthorn, ed., *The British working class novel in the twentieth century* (London: Edward Arnold, 1984), 111–24.

Simmons, J., *Sean O'Casey* (London: Macmillan, 1983).

Smith, S., *The origins of modernism: Eliot, Pound, Yeats and the rhetorics of renewal* (Hemel Hempstead: Harvester, 1994).

Stewart, A. T. Q., *The Ulster crisis: resistance to Home Rule 1912–14* (London: Faber, 1969).

Strachey, R., *The cause: a short history of the women's movement in Great Britain* (Bath: Chivers, 1974).

Sturken, M., 'The wall, the screen and the image: the Vietnam Veterans' Memorial', *Representations*, 35 (1991), 118–42.

Tate, T., *Modernism, history and the First World War* (Manchester: Manchester University Press, 1998).

Thompson, W. I., *The imagination of an insurrection: Dublin, Easter 1916: a study of an ideological movement* (Oxford: Oxford University Press, 1967).

Thomson, A., 'The Anzac myth: exploring national myth and memory in Australia' in R. Samuel and P. Thomson, eds., *The myths we live by* (London: Routledge, 1990), 73–82.

Tierney, M., Bowen, P. and Fitzpatrick, D., 'Recruiting posters' in D. Fitzpatrick, ed., *Ireland and the First World War* (Dublin: Trinity College History Workshop, 1986), 47–58.

Till, K., 'Staging the past: landscape design, cultural identity and *Erinnerungspolitik* at Berlin's Neue Wache', *Ecumene*, 6 (1999), 251–83.

Travers, P., ' "Our fenian dead": Glasnevin cemetery and the genesis of the Republican funeral', in J. Kelly and U. Mac Gearailt, eds., *Dublin and Dubliners* (Dublin: Helicon, 1990), 52–72.

Turpin, J., 'Cuchulainn lives on', *Circa* 69 (1994), 26–31.

Oliver Sheppard 1865–1941 (Dublin: Four Courts Press, 2000).

Tylee, C., *The Great War and women's consciousness: images of militarism and womanhood* (London: Macmillan, 1990).

Ulster division memorial (PRO WO 32/5868, Nos, 6, 10, 10A and 11), October–December 1919.

Vance, J. F., *Death so noble: memory, meaning and the First World War* (Vancouver: University of British Columbia Press, 1997).

Verdery, K., *The political lives of dead bodies: reburial and postsocialist change* (New York: Columbia University Press, 1999).

Wagner-Pacifini, R. and Schwartz, B., 'The Vietnam Veterans' Memorial: commemorating a difficult past', *American Journal of Sociology*, 97 (1991), 376–420.

Warner, M., *Monuments and maidens: allegory of the female form* (London: Picador, 1985).

Welch, R., ed., *The Oxford companion to Irish literature* (Oxford: Clarendon, 1996).

Whalen, R. W., *Bitter wounds: German victims of the Great War* (Ithaca: Cornell University Press, 1984).

Whelan, K., *The tree of liberty: radicalism, Catholicism and the construction of identity* (Cork: Cork University Press, 1996).

Whelan, Y., 'Sackville Street/O'Connell Street: turning space into place, the power of street nomenclature', *Baile* (Dublin: University College Dublin Geography Society, 1997), 4–10.

'Monuments, power and contested space – the iconography of Sackville Street (O'Connell Street) before Independence (1922)', *Irish Geography*, 34 (2001), 11–33.

'The construction and destruction of a colonial landscape: commemorating British monarchs in Dublin before and after Independence', *Journal of Historical Geography* (forthcoming)

Winberry, J. J., 'Symbols in the landscape: the Confederate memorial', *Pioneer America Society Transaction*, 5 (1982), 9–15.

' "Lest we forget": the Confederate monument and the southern townscape', *Southeastern Geographer*, 23 (1983), 107–21.

Winter, J., *Sites of memory, sites of mourning: the Great War in European cultural history* (Cambridge: Cambridge University Press, 1995).

Winter, J. M., *The Great War and the British people* (London: Macmillan, 1985).

Withers, C., 'Place, memory, monument: memorializing the past in contemporary highland Scotland' *Ecumene*, 3 (1996), 325–44.

Yates, F., *The art of memory* (London: Routledge, 1978).

Yeats, W. B., *The Irish Statesman*, 10, 9 June 1928.

Collected poems (London: Macmillan, 1982).

Young, J. E., *The texture of memory: holocaust memorials and meaning* (London: Yale University Press, 1993).

Zerubavel, E., *Hidden rhythms: schedules and calendars in social life* (Chicago: Chicago University Press, 1981).

Zneimer, J., *The literary vision of Liam O'Flaherty* (Syracuse: Syracuse University Press, 1970).

Newspapers and journals

An Claidheamh Soluis
Belfast Newsletter
Cork Examiner
Dublin Evening Post
Evening Herald
Freeman's Journal

Irish Builder
Irish Freedom
Irish Independent
Irish News and Belfast Morning Post
Irish Press
Irish Times
Longford Leader
Newtownards Chronicle
Sligo Independent
The Worker's Republic
United Ireland Journal

Index

Cambridge Studies in Historical Geography

*Titles marked with an asterisk * are available in paperback.*

www.ingramcontent.com/pod-product-compliance
Ingram Content Group UK Ltd.
Pitfield, Milton Keynes, MK11 3LW, UK
UKHW040705180125
453697UK00010B/429